FIGHT, FLIGHT OR FLOURISH

Endorsements

"A marvellous compilation and translational book. When we wonder about who we are or might be, we often don't understand the complex interactions of our neurological, hormonal, emotional and mental systems that foster or inhibit these realizations, hopes and fears. Ingra du Buisson-Narsai decodes and demystifies a great deal of research to date about these processes. It is a required primer for those who wish to follow and interpret emerging research discoveries and begin to understand what it means to be human."

Richard Boyatzis – Distinguished University Professor, Departments of Organizational Behavior, Psychology, and Cognitive Science, Case Western Reserve University; co-author of the international best seller, Primal Leadership, *and the new book,* Helping People Change

"An insightful exploration and explanation of human behavior through the eyes of neuroscience – a wise neuro-investment that will maximize your individual and social performance."

Dr John Demartini – International best-selling author of The Values Factor

"Are you looking at boosting your neurocapital? Then Ingra's book is a must read. She makes neuroscience – a field that can be technically daunting and overwhelming at the best of times – a no-brainer. As has been said, the highest form of sophistication is simplicity. This is exactly what Ingra does with this complex field, but she goes even further by deftly showing the practical application of neuroscience over a diverse range of domains in the world of work. This makes it very easy for one to invest in one's neurocapital. Make the investment!!"

Theo H. Veldsman – Work Psychologist; Visiting Professor in Work Psychology, University of Johannesburg; Extraordinary Professor, USB, Co-editor of Leadership Perspectives from the Front Line *and author of* Designing Fit-for-PUrpose Organisations

"From the outset, the book challenged me – then I realised it was not challenging reading, but rather it challenged some of my core beliefs. Simply put, the premise made more sense than I had hoped and was far more relevant to my day-to-day life than I would have imagined. While the concept of using integrative organisational neuroscience as a catalyst for change seems complex and complicated, the strategies outlined in the book to help individuals and organisations to flourish are accessible and practical. Thanks to the insight provided by this study, I have developed a far greater appreciation for the need to actively manage a wider range of factors that impact on my own mental wellness, particularly stress, and how I can proactively play a more positive role in my own family's lives and the lives of my colleagues.

"Ultimately, the book has provided me with some basic guidelines to help me become a better human being and a more balanced and effective CEO. I would encourage both proponents and sceptics of this illuminating subject matter to read this book."

Darren Hele – Chief Executive Officer, Famous Brands Ltd

"*Fight, Flight or Flourish* is a great read and asset for leaders who are interested in creating a flourishing environment for their organisation. I was particularly attracted by her provision of practical tools like her guidelines on, 'what to do, and what not to do, to build neural awareness and leverage NeuroCapital'. This is a gift to any leader who wants to lead consciously and inclusively."

Nene Molefi – Owner/CEO of Mandate Molefi; Consultant; Board member of The Center for Global Inclusion; Author of A Journey of Diversity & Inclusion in South Africa

"As an astute professional of applied behavioural neuroscience, Ingra is the real deal. This is well portrayed in her carefully crafted book, filled with important and insightful neuroscience informed nuggets for enhancing human capital. I highly recommend this important and most rewarding read."

Dr Etienne van der Walt – Accomplished neurologist;
co-founder and CEO of Neurozone Pty Limited

"The use of concepts in neuroscience in a number of fields in education, leadership and coaching is increasing. These often present a very partial or simplistic account of the science. They often over claim for the science and perpetuate neuro-myths. Rightly, many have become wary of the claims. The author seeks to overcome this simplification by taking the reader through what is known but also noting the limitations. She succeeds in presenting a clear account of current understanding and using that to create practical interventions that can be used and are based in thinking from neuroscience. The book is a starting point for a journey into this field that will repay careful reading and consequently makes a useful contribution to the application of neuroscience to practice."

Prof David Lane – Director Professional Development Foundation (PDF);
Fellow of the Academy of Social Sciences; Honoured for Distinguished Contribution to
Professional Psychology by the British Psychological Society

"In this enlightening book, Ingra has succinctly brought the quite complex field of neuroscience to bear on the world of work and organisational leadership. Drawing on key neuroscientific principles, concepts and techniques, she has practically demonstrated how the 'hands-on' application thereof can lead to a greater understanding and improvement of leader, team and organisational behaviour and performance."

Dr Chris van der Burgh – Former Chief, Division for Operations, United Nations Office on
Drugs and Crime; current Director, NeuroCapital Coaching and Consulting

"I had the privilege of reading *Fight, Flight or Flourish* by Ingra du Buisson-Narsai before it was published. The world has been waiting for this book. The value of this book to individuals, the world of business, and intellectuals is the comprehensive and straightforward way Ingra has taken the reader on a journey of unlocking human capital. She explains how the brain works understandably. Included are helpful guidelines about handling stress, the importance of sleep, the value of stillness, the importance of emotional regulation, the value of social connectivity, and the power of goals. I will highly recommend this book to family, friends, and clients."

Ilze Alberts – Psychologist; Speaker; Author of Passing the Torch

"The aim of this publication is to introduce neuroscience to the workplace. Ingra is commended for her brave efforts to explain difficult neuroscientific concepts in a language that will appeal to readers. The strategies discussed in the book will contribute positively to the self-development of leaders and employees alike."

Dirk Geldenhuys – Professor of Industrial and Organisational
Psychology, University of South Africa

"If you are fascinated by human behaviour, Ingra du Buisson-Narsai's book, *Fight Flight or Flourish*, is a must read. It is the most comprehensive and engaging resource I have read on the topic of neuroscience. The book is filled with practical tips and tools to unlock human capital. I see enormous value in using this practical guide – for leaders, teams and organisations."

Niel Steinman – CEO, Peoples Dynamic Development

"Ingra du Buisson takes the complex topic of neuroscience and leads the reader gently through the concepts that underlie the point of the book to enable flourishing. Based on academic literature, Ingra assimilates theory into real-life actions and behaviours, giving examples that everyone can relate to. Then she turns it all into actionable steps through which individuals can build the courage to break away from immobilisation to fulfilment. Thank you, Ingra!"

Terri Carmichael – Associate Professor of Management
Education, Wits Business School

"Life in the time of disruption, (now), is like walking across the thin, ice surface of a lake in the dark! Ingra's collection of fresh and practical ideas based on neuro-evidence is like finding a map after being lost in the wild. Affirming and priceless. A book for every will-be unleasher of human potential!"

Derek John Jooste – Organisational Change Consultant par excellance with Snaptech
International and founder of Kismet Human Potential

"In her insightful and practical book on organisational neuroscience, Ingra du Buisson takes us into the world of neuroscience to understand the way our brain sees the world and how it governs our 'behaviours, decisions and reactions'. In her imminently accessible narrative, she delves into the new discipline of 'brain-based leadership' through her empirical research and studies. Ingra takes us on a journey to help us find ourselves in our search for personal meaning and purpose – and more particularly to understand how we are leading ourselves and others.

"She starts with the fundamental importance of the unspoken question posed by the brain stem, 'How do I get to physically survive?', before identifying the unspoken questions of life from pre-school years, to questions about whether each of us is actually in charge of our brain, or if our brain is in charge of us. She answers questions about the emotional brain, the impact of stress on the brain and how the brain deteriorates, and discusses the need to still the mind to leverage the 'sub-conscious mind and all its superpowers'.

"Looking through the lens of neuroscience, Ingra leads us to a deeper level of understanding about human thinking and behaviour. She explores the critical skill of attention to help us become 'the architects of our lives' in a very real sense. The brain's ability to rewire itself (neuro-plasticity) helps us to develop memory and learn new habits. Giving us seven habits to develop 'NeuroCapital', she invites us to develop new habits and new behaviours, at the same time as learning to 'stress right', with the ability to develop resilience. Sharing with us the dangers of a lack of sleep, to 'brain flossing' through mindfulness, she helps us understand how emotions are made and our limitless ability to develop emotional fluency to create that upward spiral of fluency. Whether you want to develop neuro-insight, build trust in the workplace, reframe your leadership strategies or social connections at work or at home, develop an ability to show empathy – growing yourself, your team or your organisation – or develop strategies for goal setting, Ingra unlocks the secret of creating 'genuine, lasting change in our lives'. An important book to take us beyond survival to be able to thrive in our complex, ever-changing world. Not to be missed."

Sunny Stout-Rostron – DProf (Faculty, School of Business at the
University of Stellenbosch); Director, People Quotient Pty Ltd (PQ); Fellow,
IOC at McLean Hospital, Harvard Medical School Affiliate; author of Business Coaching:
Wisdom and Practice *and* Leadership Coaching for Results: Cutting-edge practices for
coach and client

"The general (or organisational) utility of knowledge generation is well expressed in the fundamental premise of Chris Argyris' *Action Science*, i.e. that knowledge must be actionable. This, I believe, Ingra has achieved successfully in her book by taking complex neuroscience theory ('there's nothing more practical than a good theory' – Kurt Lewin), and through her extensive organisational (leadership) development experiences (many of which I collaborated with her on), turned it into actionable practices of NeuroCapital in the workplace. A case in point is that while we strive for higher order thinking in the workplace, a simple, scientifically-based neuro-insight that safety comes first (lower order functioning, and expressed, for example, in people experiencing organisations as psychological unsafe) will derail all our efforts to help employees think and do rationally (higher order). This is essential reading for the serious organisational development specialist; the VUCA of the new world of work demands this."

Dr Andrew Johnson – Chief Learning Officer at Eskom; intrepid Action Scientist; Co-editor of Leadership Perspectives from the Front Line

The book is highly intellectual, and contains all the supporting facts and rationale that will attract the reader who is searching for more depth and a better understanding of Neuro Capital studies. And yet, for the hands-on, strapped for time, on the move, "give me the topline only" readers (much like myself), the astute summaries at the end of each chapter, as well as the engaging models, examples and research statistics, proved to be extremely beneficial and an interesting read. I found the insights in this book most valuable, and know I will refer back to it in my daily work.

Val Nichas, VBN Consultants

"Ingra's passion for making neuroscientific application easily digestible for the mainstream reader is an effective and refreshing foundation for the book. Everything from the basic biological principles of neural functioning to the complex social dynamics inherent to the interconnected team, to the mitigation of psychological biases. Her use of simple models aids the reader in creating practical applications to enhance personal and professional performance."

Dr Dan Radecki – Co-founder, Academy of Brain-based Leadership (ABL)

"This book represents Ingra Du Buisson-Narsai's brave, relevant and highly informative explorational work in neuroscience, as a fast-developing field of study and practice and application in all fields of human functioning. She presents theory and practice in a logical, comprehensive and authorised manner. The most outstanding contribution is the integration of the knowledge and how to use this practically in personal as well as organisational and professional scenarios. This is a must read and user-friendly application suggested for all organisational psychologists, consultants, coaches, leaders and students."

Frans Cilliers – Psychologist; Emeritus/Extraordinary Professor, Organisational Psychology

First published in 2020.

ISBN: 978-1-86922-828-6
eISBN: 978-1-86922-829-3 (ePDF)

Published by KR Publishing
P O Box 3954
Randburg
2125
Republic of South Africa

Tel: (011) 706-6009
Fax: (011) 706-1127
E-mail: orders@knowres.co.za
Website: www.kr.co.za

Typesetting, layout and design: Cia Joubert, cia@knowres.co.za
Cover design: Marlene de'Lorme, marlene@knowres.co.za
Editing and proofreading: Jennifer Renton, jenniferrenton@live.co.za
Project management: Cia Joubert, cia@knowres.co.za

FIGHT, FLIGHT OR FLOURISH

HOW NEUROSCIENCE CAN UNLOCK HUMAN POTENTIAL

Ingra du Buisson-Narsai

kr
publishing

2020

Acknowledgements

The inspiration for this book is built firmly on the shoulders of my family, friends, teachers, clients, colleagues, and authors who have shaped me personally and professionally.

But moving from inspiration to achieving a main goal demands a lot of perspiration and persistence. Why I was driven to write this book can be captured in the following quote (sent to me by my dad):

> *"There is only one way: Go within. Search for the cause, find the impetus that bids you write. Put it to this test: Does it stretch out its roots in the deepest place of your heart? Can you avow that you would die if you were forbidden to write? Above all, in the most silent hour of your night, ask yourself this: Must I write? Dig deep into yourself for a true answer. And if it should ring its assent, if you can confidently meet this serious question with a simple, "I must," then build your life upon it. It has become your necessity. Your life, in even the most mundane and least significant hour, must become a sign, a testimony to this urge."*
>
> Rainer Maria Rilke, Letters to a Young Poet[1]

So many people contributed ideas through formal assignments, projects, communities of practice, critiques and formal and informal conversations. In a sense, this book is therefore collaboration with all of them. It indeed takes a village to write a book. The following are a few members of this village, to whom special thanks are due:

Teachers and lecturers, some of whose valuable contributions are specifically cited in this book: Dan Radecki, David Lane, Etienne van der Walt, Frans Cilliers, Golnaz Tabibnia, Ian Weinberg, Ilze Alberts, John Demartini, Matthew Lieberman, Richard Boyatzis, Simon Whitesman, Susan Kriegler, and Vasi van Deventer.

Dear colleagues who participated (and some who still do) in many hours of study groups or communities of practice: Adelle Bester, Anize van Zyl, Beth Norden, Carla Street, Estelle Coetzer, Iftikar Nadeem, Kay Waldron, Linda Hayes, Lynn Andries, Samad Aidane, and Scott Halford. Thank you for all the enticing discussions and friendships made.

Many generous and insightful colleagues across diverse fields who offered me their insights. I benefited greatly from your valuable comments, collaborations on projects, formal critiques, informative conversations, joint presentations at conferences and friendships:

Adam Martin, Adelaide McKelvey, Andrew Johnson, Andrew Morris, Anita Arendse, Anja van Beek, Arlene Botha, Bronwyn Bainbridge, Busie Mjimba, Cherylene de Jager, Clint Ducie, Darren Hele, Derek Jooste, Dirk Geldenhuys, Francois Hugo, Gail Wrogeman, Gerrit Walters, Gillian Contry Taylor, Grace Chang, Hope Lokoto, Innette Taylor, Johan Olwagen, Kevin Hedderwick, Lebo Makgabo, Leigh Ann Crane Silber, Linda Sinclair, Magda Ross, Marina Gunther, Marissa Brouwers, Markus Moses, Melanie Hall, Mitzie Hollander, Natalie Cunningham, Natasha Winkler-Titus, Nene Molefi, Niel Steinman, Phil Dixon, Pratiba Daya, Regina Calitz, Reinette Nel, Rob Reiche, Rob Jardine, Rose Pillay, Sandra Meso, Sandri van Wyk, Stefan Botha, Steven Breger, Sunny Stout Rostron, Tanya Maddi, Terri Hunter, Terry Boardman, Theo Veldsman, Theofanis Halamandaris, Val Nichas, Varsha Naran and Xander van Lill.

Nokuthula Mnguni for faithfully working with me when energy plunged, hard disks failed, and ideas ran dry.

Knowledge Resources for editing and publishing this book, and particularly to Wilhelm and Cia, for sculpting the book's message in meaningful ways, thus making the writing journey less daunting.

My deepest appreciation is due to my extended family for the lived experiences of fight, flight or flourish. My dear parents, Chris and Betsie, thank you for your limitless support from the outset and for taking me to a keynote address by Victor Frankl when I was 14 years old, which ignited the flame for my own search for personal meaning and purpose. Thank you also to my dad for the initial proofreading of the manuscript (proofreading at its best).

Finally, I wish to thank Yatin, my husband, for challenging my thinking, and helping me bring down my ideas to earth over a strong espresso and with intellectual insight. You are a force. And to my daughters, Mira and Tara – I owe a debt of gratitude to you (which I can never repay) for tolerating the long hours and compelling me to practice what I preach. I hold you and your unique talents in high esteem.

To all those who hunger for what is missing

Table of contents

List of figures

List of tables

About the author

Ingra is the co-founder and Director of NeuroCapital Consulting, which consults to some of South Africa's leading and most admired companies. She has 20 years of executive level experience in corporate South Africa and is a Registered Organisational Psychologist in private practice. Ingra is a frequent best practice "Sharer" at conferences, business schools and in the media. Her unique contribution is as a catalyst for change, using integrative organisational neuroscience, leadership coaching, and crucial mentoring conversations as change drivers.

Ingra is also an established leadership and executive coach and is affiliated with the Wits Business School (WBS) in South Africa, where she supervises and examines the work of post-graduate students in Business and Executive Coaching.

Ingra is an Executive Committee Member of the Society for Industrial and Organisational Psychology of South Africa (SIOPSA) and the Chair for the Interest group of Applied Organisational Neuroscience (AONS).

Ingra's academic qualifications include an MCom (Organisational Psychology), a PGCNL (Neuroleadership), an MSc in Professional Development (Neuroscience of Leadership), and a Ph.D. in Organisational Neuroscience (in process).

She has also published a chapter on Neuroscience-Based Leadership in a scientific peer-reviewed book on Leadership (2016). Ingra actively pursues the increasing visibility of neuroscientific methods and diagnostics in the study of organisational behaviour.

Acronyms

ACC	Anterior Cingulate Cortex
ACTH	Adrenocorticotrophic Hormone
ADHD	Attention Deficit Hyperactivity Disorder
AI	Anterior Insula
BDNF	Brain Derived Neurotrophic Factor
CEN	Central Executive Network
CRF	Cortico-Releasing Factor
DLPFC	Dorsolateral Prefrontal Cortex
DMN	Default Mode Network
DNW	Default Network
ED	Executive Director
EEG	Electroencephalogram
EFT	Episodic Future Thinking
HRV	Heart Rate Variability
MNS	Mirror Neuron System
MPFC	Medial Prefrontal Cortex
NEA	Negative Emotional Attractors
PEA	Positive Emotional Attractors
PFC	Prefrontal Cortex
PNI	Psycho-neuroimmunology
PNS	Parasympathetic Nervous System
PPC	Posterior Cingulate Cortex
REM	Rapid Eye Movement
SN	Salience Network
SNS	Sympathetic Nervous System
SWS	Slow Wave Sleep
TPN	Task-Positive Network
VLMPFC	Ventrolateral Medial Prefrontal Cortex
VMPFC	Ventromedial Prefrontal Cortex

Terms for, and directions of, brain regions:

Anterior	Front
Posterior	Back
Dorsal	Above
Ventral	Below
Medial	Centre
Lateral	Left and Right

Icons used

Stress: manage it and build resilience

Sleep health

Stillness of mind: switch to mindful awareness

States of mood

Simple fluency and regulation of emotions

Social connectivity

Stratification of purpose and goals – the will, the way and the habit

#Neuro-hacks

The aims and objectives of this book

This book does not purport to be the "final word" or blueprint on how neuroscience can be applied in the workplace, contribute to outperformance, engagement, flourishing or agility at work.

Instead, the book seeks to introduce some of the key neuroscientific concepts and principles that underlie 'how' the brain sees our world and 'why' it reacts the way it does in governing our behaviours at work. It is based on both theoretical and empirical evidence. Trying not to be too academic or ponderous, the evidence and ideas are new and unexpected.

Drawing on learnings from interdisciplinary research in psychology, neuroscience and evolutionary biology, as well as experience in formal organisations, *Fight, Flight or Flourish* offers some surprising strategies to enable you to flex from fragile behaviour and fatigued thinking to clear, focused thinking and flourishing behaviour at work.

Flourishing is described in the APA Dictionary of Psychology[2] as a condition denoting good mental and physical health: the state of being free from illness and distress but, more importantly, of being filled with vitality and functioning well in one's personal and social life. It requires embracing both sides (the good and the bad) and linking those to our personal meaning and the purpose that we bring to the world. This represents a NeuroCapital Balance Sheet.

What is written about can be applied at any level. Much of it can be applied to yourself, as well as to leadership, teams and organisations. Running throughout the book are ideas on what you can actually do to cultivate each strategy and reflect about what is likely to work or not.

Finally, a little note about grammar. I did not always stick to the rules. I have done this to make it an easier read and hope that it will be understood as such.

1

There is no learning without having to pose a question

"We absolutely must leave room for doubt or there is no progress and there is no learning. There is no learning without having to pose a question. And a question requires doubt. People search for certainty. But there is no certainty. People are terrified – how can you live and not know? It is not odd at all. You only think you know, as a matter of fact. And most of your actions are based on incomplete knowledge and you really don't know what it is all about, or what the purpose of the world is, or know a great deal of other things. It is possible to live and not know."

Richard P. Feynman[3]

In this book, I am heeding Feynman's[4] call as set out above, which was captured in *The pleasure of finding things out;* namely, that it is ok not to 'know' completely but still we won't stop asking. I ask what science can tell us about what "neurally aware" corporate citizens "do" to help business and people thrive.

This book introduces some of the key concepts which underlie 'how' the brain sees our world and 'why' it reacts the way it does in governing our behaviours, decisions and reactions. Since we all have a brain, this information is universally relevant and can help us understand how to better manage the limitations inherent in a brain that has not evolved at the rapid pace at which our society has evolved. Importantly, we also focus on possible practical applications derived from this information, such as, 'What tangible actions can I take in order to lead in a truly 'brain-friendly' or neurally-aware manner?'

As technology continues to evolve, neuroscience is able to become more precise in its quest to understand the biology behind human behaviour. The fields of organisational behaviour and leadership development have been reinvigorated with the potential applications coming out of the field of neuroscience. The fact that leadership has a broad definition (for example, parent, teacher, coach, among others) makes the brain-based leadership space a discipline of great public interest and numerous possible applications, because we are all, to some extent, leaders.

Supercharged at work

In today's supercharged world at work, we are grappling with the need to learn new ways of leading ourselves and leading others: how to manage people well who are not sitting in the office where you can see them; how to measure and reward in a virtual workplace; how to virtually connect diverse skill sets that will lead to optimum results; and how at the same time to be nice to others at work. Most of us vastly underestimate the need for belonging, personal meaning and purpose in this supercharged world.

As noted by Professor Klaus Schwab[5] in an address at the World Economic Forum's annual meeting, "We need leaders who are emotionally intelligent and able to model and champion co-operative working. They'll coach, rather than command; they'll be driven by empathy, not ego. The digital revolution needs a different, more human kind of leadership" (read corporate citizens). This is easier said than done.

Alternatively put, we need fearless organisations where candor and openness are welcomed through psychological safety. According to Edmondson[6], safety takes off the brakes that keep people from achieving what is possible. Leaders have two vital tasks: to build psychological safety to spur learning and avoid preventable failures; and to set high standards and inspire and enable people to reach them. In other words, today's leaders must motivate people to do their very best work by inspiring them, coaching them, providing feedback, and making excellence a rewarding experience.

2

A brainwise value chain – NeuroCapital drives human capital

I propose that neuroscience can offer some more in-depth insights into how to live in a supercharged world. Neuroscience is a field of study that includes a number of different disciplines (life and natural sciences) that deal with the structure, development, function, chemistry, pharmacology and pathology of the nervous system. Leveraged technological breakthroughs in brain imaging and computational modelling illuminate the inner workings of the human brain and suggest that many brain processes have evolved over millennia for specific evolutionary aims – often with a high level of automatism.

The applied field of organisational neuroscience helps us to understand behaviour at work and is constructed as a multi-disciplinary topic, drawing from neuroscience and other social sciences that use neuroscientific methods, for example, work psychology, social psychology, neuropsychology and evolutionary psychology. To expand on this let's use evolutionary psychology, which underlines the fact that human beings are hardwired to practice certain habits and "holds that although human beings today inhabit a thoroughly modern world of space exploration and virtual realities, they do so with the ingrained mentality of Stone Age Hunter-gatherers. You can take the person out of the Stone Age but you cannot take the Stone Age out of the person".[7] In fact, some of the functions of the brain specialise over time to respond to environmental constraints.[8]

The focus of this book is NeuroCapital – the underpinning resource that enables the effective use of human thinking and behaviour. Neuroscience may give us a deeper level of analysis in the study of our work behaviour. Figure 1 below shows the components of a NeuroCapital value chain.

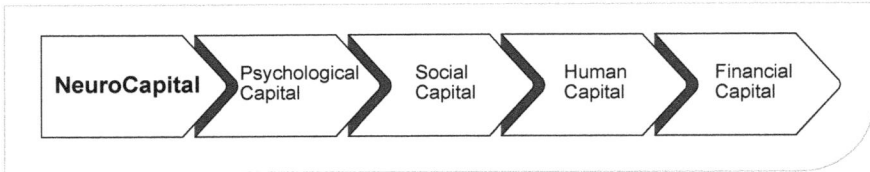

Figure 1: NeuroCapital value chain

NeuroCapital, which sounds like a brain ledger or profit and loss statement, can, in fact, be measured. In this book, I am aiming to simplify complex ideas and concepts and translate them into meaningful practical action (by providing reflective questions and 'neuro-hacks'). The question that I aim to answer is: If you know how your brain works, what will you 'do' or 'not do'? To put it in a more neuroscientific way, building your NeuroCapital means deliberately applying the organising principle of the human brain to minimise danger, maximise reward and deliberately implement proven and interrelated strategies to build a life of flourishing (neural thriving). Another way to think of this (although this is not explored in detail in this book) is to develop 'in zone performance' underpinned by neurobiological proficiencies. In Zone Performance is the state when your nonconscious automatic brain confidently governs your conscious thinking, permitting peak performance in the moment.[9] This is done through objective brain assessments and brain training Apps. For now, let's focus on foundational concepts and simple, albeit science-based, strategies.

At a glance

Following are the topics that are discussed next, as well as an outline of the design template for each chapter that will serve as a reference to the journey you will need to embark upon to build NeuroCapital at work:

- Start with the brain in mind.

- The Big 5 foundational concepts.

- From scanner to boardroom and beyond.

- Seven science-based strategies to shift from fight, flight and freeze to flourish.

- Neuro-Insights – short reflective practice exercises to slow down your thinking, learn more and create lasting change. Slowing down our thinking deliberately through self-reflection or a reflective conversation can develop new neural pathways.

  The development of reflective practices is underpinned by an ability and willingness to question and explore ways of acting and thinking as we engage in business activities. The practice of reflecting, in and on action, makes it possible to change our current understanding of action by framing the issue or encounter in a different or novel way, or by improvising new ways to solve the issue at hand.[10]

- Neuro-hacks – what to do, and what not to do, to build neural awareness and leverage NeuroCapital.

- A Neuropedia of helpful definitions and neuroscience terms is also supplied at the end of the book.

3

Making it as simple as possible

start with a scientific perspective, trying, as Einstein advised, to "make everything as simple as possible – but not more so".[11] The challenge with applying neuroscience or brain data to organisational behaviour is that an oversimplification of the proclaimed neural underpinnings of behaviour can render the study of anything 'neuro' as a short-lived movement.[12] So, let's not ride the hype-wave – formal organisations are far from using neuro-imaging techniques to gauge specific behaviour. Scanning has its limits, and it cannot tell us who the right CEO is to turn our struggling company around. (Well, not this year, nor the year after.)

The pitfalls of brain-based approaches are that they are seen by some as reductionist; research findings are often preliminary, there is a lack of replication studies, and research samples are typically too small to generalise findings.[13] The most effective forms of engagement within the field of neuroscience by organisational scholars are advocated as follows by Van Ommen and van Deventer[14]: "…we find ourselves in the midst of a proliferation of neuro-centred disciplines, establishing what some call a neuro-culture." The authors warn against neo-liberal (i.e. new and faddish) approaches to studying human behaviour, as these approaches lack scientific rigour, replication studies, and are driven by populist anecdotal interpretations and broad generalisations.

So, if you put a picture of an MRI in your PowerPoint, people might be more willing to accept your content, but an over reliance on neuro-centred disciplines where we increasingly understand, speak about, and act upon ourselves and others as biological beings, is an oversimplification of our personhood.

The applied neuroscience approach (adopted in this book) is to simplify cutting-edge research and to make it accessible to those fascinated by human behaviour. To this end, neuroscience studies will be described in language that is clear and linked conceptually to the human behaviour to which it is relevant.

4

It's not rocket science... it's neuroscience

4.1 Start with the brain in mind

Although we are far away from a unified theory about how the brain functions, some themes come up repeatedly to inform us how to engage with the topic using brain science as a lens to understand behaviour at work:

"No one said this was easy. But the subject matters."

Robert Sapolsky[15]

The main themes that appear to be repeated in the neuroscience literature include the below:

- The context and meaning of a behaviour are usually more interesting and complex than the mechanics of the behaviour. (Examples are team dynamics, organisational culture, family dynamics, the traffic on the day, the boss' mood, etc.)

- To understand things, you must incorporate neurons, hormones and early development and genes.

- There are not separate categories – there are few clear-cut causal agents, so do not count on there being the brain region, the neurotransmitters, the gene, the culture influence, or the single anything that explains a behaviour.

- Instead of cause, biology is repeatedly about propensities, potentials, vulnerabilities, predispositions, interactions, modulations, contingencies, if/then clauses, contexts, dependencies, and the exacerbation or diminution of pre-existing tendencies. Biology works in circles and loops and spirals.

"It has been calculated that there are more possible connections in one human brain than there are atoms in the universe."

Unknown

The brain is comprised of approximately 100 billion neurons. These neurons make up the basic building blocks of the brain and are commonly known as grey matter. The brain also contains billions of nerve fibres called axons and dendrites, which are identified as white matter. The white matter is white due to the fatty substance (myelin) that surrounds the nerve fibres (axons). The neurons are linked together by trillions of connections (or more) called synapses. These synapses are minuscule spaces between sending and receiving neurons, called synaptic clefts.

Brain functioning is the result of neurons firing and triggering (or activating) each other by using an electrochemical process, all the while continually monitoring and modifying these connections to maintain optimal processing.[16]

Various thinking frameworks and explanatory models exist to explain brain functioning. (See Appendix A for a discussion of the thinking frameworks that aim to articulate whole-brain functioning that can be applied to workplace behaviour.) Various levels of analysis enable us to peer into behaviour at work, but for now, I am keeping it simple. Please be aware that the kinds of operational – i.e. useful – explanations that follow are necessarily inaccurate, due to oversimplification.

4.2 Three metaphorical (but not literal) layers

I use the triune brain hypothesis by Dr Paul Maclean, a neuroscientist and psychiatrist who developed the fascinating theory of the "triune brain" to describe its evolution and to try to reconcile rational human behaviour with its more primal origins.[17] The triune brain model has now been replaced with more sophisticated models, but it provides a foundational knowledge in that it shows that there is a sequence to brain development. This sequence has a significant effect on neural firing and information flow in the brain, as well as giving insight into the manifestation of mental health and pathogenic conditions, such as narcissism, schizophrenia and other mental disorders.

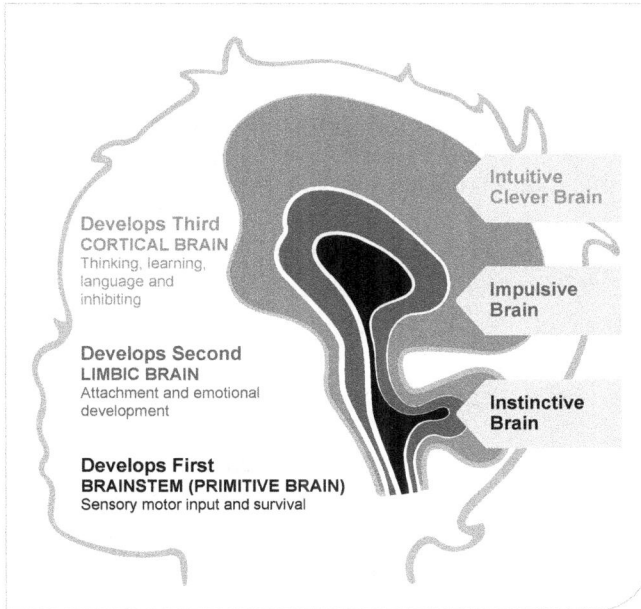

Figure 2: Our three brains (Source: Adapted from MacLean)18

Layer 3: The cortical brain or clever/intuitive brain

The 'you' that 'knows better' is located in the cortex. This is the part of the human brain that is much larger and vastly better developed than in animals. This is your conscious mind; it is responsible for logical thinking, reasoning ability, long-term planning, understanding, wisdom, self-governance, and enlightenment. This part of the brain is the intelligent computer, capable of complex problem-solving and cleverness, and it is also the data bank for facts, figures, principles and theories. (All of our data are not conscious – i.e. 'on the screen' – at the same time, so a great part of it is pre- or subconscious. For the sake of simplicity, we will call this part of you the human brain or the conscious mind.)

This part of the brain is linked to the outside world through the senses, i.e. vision, touch, hearing, taste and smell. Information from the sensory world is processed and filtered through it to the lower brain, where it is emotionally processed and stored. In their turn, electrochemical signals in the form of emotions and body sensations from the lower brain and the body are filtered through to the conscious mind, where they appear on the screen of awareness if they are strong enough.

This intelligent conscious mind has all the information, but has little control over the impulsive emotional and instinctual unconscious mind where feeling and behaviour are controlled. Sometimes the emotions and behaviours surging up from the lower brain can be so powerful that the clever brain is bypassed, short-circuited or hijacked. That's when 'everything goes blank' and you only realise – and regret – what you have said or done after the emotional storm has abated.

Layer 2: The limbic system or emotional/impulsive brain

Emotions are generalised in the limbic system, which is a cluster of structures that lies beneath the cortex. The system evolved very early in mammalian history. In humans, it is closely connected with the more recently evolved cortical areas. The two-way traffic between the limbic system and the cortex allows emotions to be consciously felt and conscious thoughts to affect emotions.[19]

Emotions are constructed by different networks of brain parts, including the hypothalamus and pituitary gland. These are the conductors of the symphony of chemical molecules that produce physical reactions, such as an increased heart rate and muscle contractions, and experiences, such as your feelings and emotions.

Conscious emotion

Emotions are mostly produced in the limbic system, which does not support consciousness itself. Intense emotions create 'knock-on' activities in the cortex, especially in the frontal lobes, which we experience as a conscious 'feeling' or mood. Sometimes an emotion is clearly linked to an experience. At other times the cause is not apparent, but being aware of the emotion makes it easier to understand what is happening to us.[20]

Unconscious emotion

We have developed a conscious emotional system, but we keep the primitive, involuntary responses at the heart of emotion. A frightening sight or sound, for example, registers in the amygdala before we are even aware of it. While the sensory information is being sent to the cortex to be made conscious, the amygdala sends messages to the hypothalamus, which prompts changes that ready the body for flight, fight or appeasement. This super-fast route allows us to take instant action to save ourselves. When we get startled by a loud noise, then relax upon realising that it is harmless, we are experiencing both stages – unconscious reaction and conscious response.[21]

Emotional sponges

The emotional part of the brain opens its doors to information when the brain stem closes its doors – when you are about two years old. The doors to the limbic system remain wide open until you are about six years old. This incredible emotional and social receptivity is why there is nothing more enchanting than a toddler from about two years on. This is when you really open up to people, in order to learn everything there is to learn about emotional/social survival. This is when a baby becomes a person. Furthermore, adults react very powerfully to a young child. Their feedback – their attention, smiles, smacks, hugs, and words – are all absorbed by the darling little sponge they call their child. Based on the emotional intelligence or emotional stupidity we are surrounded by as toddlers, we develop an emotional language of being in the world. The emotional data and information are absorbed and fundamentally shape the budding personality. The unspoken question posed by the brain stem is, "How do I get to physically survive?"

According to Kriegler[22], the more complex unspoken questions asked in the pre-school years by the highly receptive limbic system are:

• "How do I need to behave to get the security, control, and freedom I need to survive emotionally?"

• "How do I get acceptance, approval, validation and/or attention?"

Through millions of interactions between caretakers and children, the blueprints are engraved. Of course, these impressions are shaped by the earliest pre-linguistic/ pre-verbal imprints of the first two years. The little bundle of need, energy and will is relentlessly shaped by positive and negative feedback, and now language is rapidly acquired. The foundations of the future conscious personality and identity are built from the emotional survival strategies that 'work'.[23]

Most of our problems, anxieties, depressions and obsessions originate here in the limbic system (see Figure 3). This is where instinct and childhood learning blend, simmer, bubble and boil in the cauldron of the unconscious mind, creating emotional and behaviour patterns that no longer serve us and which baffle the adult mind.

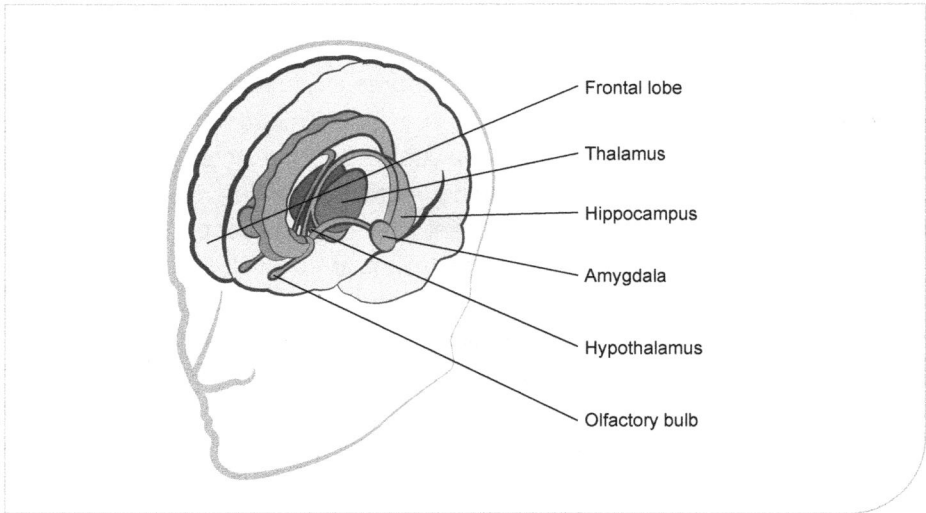

Figure 3: Layer 2 – The brain's limbic system (Source: Griffith)[24]

Layer 1: The brain stem or survival/instinctive brain

In comparison to the more intelligent, but relatively slow, 'higher' mind, this is the fast 'lower' part of the brain (also called the 'old' brain). It consists of two parts – the 'reptilian' brain and the 'mammalian' brain – so-called because our development from foetus in the womb to maturity echoes the evolutionary development of a species. The brain stem at the top of the spine is the 'oldest' and most primitive part of the brain. It is called the reptilian brain because it is responsible for our instinctual urges, body processes and physical survival. This is the part of the brain that connects us to snakes, lizards, and the T-Rex. This old, reptilian part of our nervous system is adapted to a solitary, risk-filled world – reptiles are, after all, loners – so that part of us is on the constant lookout for perceived threats.

The brain stem is where the main switch of the central nervous system sits. This is the on/off lever for fight/flight/freeze/fudge; it classifies everything that filters through to it from the body below and from the world above via the cortex in simple black and white terms. Is this safe or dangerous/pleasurable or painful? Does this serve or threaten my survival? If it is safe, the green button goes 'on'. If it is not safe, the red button is activated. The brain stem is programmed from the beginning of your existence in the womb up to about two years old. By the time you begin to walk and talk, it already 'knows' what it needs to know to ensure your physical survival, e.g. fight/be nice/be sick/cry/be clever/control.

The scary thing is that, although you may never know about or remember the exact circumstances where these far-reaching blueprints were formed, they continue to be in control of your body, your feelings and your unconscious choices. In other words, on a very rudimentary level, your life is controlled and directed by a part of your mind that will never be older than a toddler.

The conclusions reached and the decisions made by the infant that you were are still running your life. That is why it is literally true that as an adult, you do not know why you feel and behave the way you do. The default is for the impulsive and instinctive brain to take control of the intuitive brain, and our behaviour is mostly based on these blueprints or the circuit of emotional residues in the limbic system, and from the freeze-flight primitive survival mode. No matter how many times you have apologised for your behaviour, or how determined you are to not repeat the same irrational response, the same trigger automatically activates the baby blueprint circuit against your will. You need to cultivate your clever brain to avoid the trigger or reframe the trigger, not allowing it to activate the baby blueprint circuit.

In the words of Paul MacLean, writing in *The New York Times* in 1971[25] when he was gauging the difficulties that go with intolerance and violence worldwide, "…language barriers among nations present great obstacles… But the greatest language barrier," he concluded, "lies between man and his animal brain; the neural machinery does not exist for intercommunication in verbal terms".

As our understanding of the brain continues to evolve, many new technologies and learnings are available that can literally help us to change the brain for the better and to do this intercommunication a bit more easily.

4.3 The nervous system: fight/flight/freeze/fudge and friends

Running in the background: Your nervous system

Our nervous system is always running in the background, controlling our body's functions so we can think about other things – like what kind of take-away lunch we'd like to order, or how to get that proposal approved by a client. The whole nervous system works in tandem with the brain and can take over our emotional experience, even if we do not want it to.

The human nervous system is comprised of the brain, spinal cord, and all the neurons that run through the body, plus the recently acknowledged enteric nervous system. We know the gut-wrenching feeling of being rejected or having "butterflies in your stomach" before an important presentation. The gut (the enteric nervous system) is also called the second brain; it is a web-like network of neurons that control the function of the entire gastrointestinal tract. From the perspective of evolution and natural selection, the primary two jobs of our brain and nervous system over the millennia were survival and reproducing the species – to live long enough to ensure we passed on our genes to the next generation. Those humans who had nervous systems that were very swift to respond to signs of danger or life threats were the ones who lived to pass on their genes. This response is great for the survival of your genes, it is also the source of much of your struggle, discontent and pain.[26]

The autonomic nervous system

One part of your nervous system is the autonomic nervous system. This is the name given to the part of your nervous system that automatically controls activities that are crucial for your survival, such as breathing, heartbeat, blood pressure, digestion and body temperature. This system also controls your defensive behaviours, such as fight-or-flight reactions. We used to think that the autonomic nervous system was a two-gear system only, with a sympathetic gear for fight or flight, called "the stress response", and the parasympathetic gear for rest, repair and digestion, called "the relaxation response". In this two-gear system, one gear increased your activation and the other gear reduced your activation. However, that all changed with the introduction of Porges' Polyvagal Theory.[27]

Neuroception

Without us knowing it consciously, our nervous system is continuously scanning and sensing whether people, things, places or situations are safe, dangerous or life-threatening. According to the Polyvagal Theory, there is a distinct difference between the **conscious process of perception** – for instance, I am aware of the keyboard while typing a document – and the constant automatic scanning and sensing for danger or threat that occurs **outside of conscious awareness**. Porges termed this subconscious, automatic sensing process "neuroception".[28] Depending upon what your neuroception detects, it will decide what state you are in: threat detection and threat response, or feeling safe and connected.

Once your neuroception detects danger or threat, your autonomic nervous system springs into sympathetic action to respond to the situation. You either get spiraled into fight or flight, or you plunge into shutdown and immobilisation. This happens swiftly – often before you are even conscious of what the threat or danger might be.

Sometimes, your neuroception will detect danger and a lack of safety in situations that are actually relatively safe from a survival viewpoint. This faulty neuroception happens more often than you may think, due to being in a constant hyper-vigilant state. Many psychological issues such as anxiety, depression, impulsivity and addictions may be the result of faulty neuroception.[29] Faulty neuroception can be biased toward detecting danger even when there is no real danger, thus the more you improve the accuracy of your neuroception, the more you will reduce the problems that arise from it. Practicing stillness of mind and taking purposeful pauses are some specific practices to help increase the accuracy of your neuroception, which will enable a shift from a chronic survival response activation.

Three distinct branches to your survival system

Unlike the two-gear survival system, Porges[30] found that there are three separate branches and functions to the autonomic nervous system. It works as follows: in response to perceived danger, we will shift into the hyperaroused state of fight-or-flight response, and in response to a life threat, we will shift into the hypo-aroused shutdown response (appeasement). In contrast, when we are feeling safe, we shift into the vital third branch of the autonomic nervous system that Porges calls the social engagement system. To better appreciate the importance of the social engagement system, it is helpful to see how each system evolved and how they help us survive.

Evolutionary hierarchy of survival responses

The historical evolution and hierarchy of these three branches of the autonomic nervous system will help you track, befriend and work with (rather than fight against) your nervous system. See Figure 4 for the three branches of the autonomic nervous system.

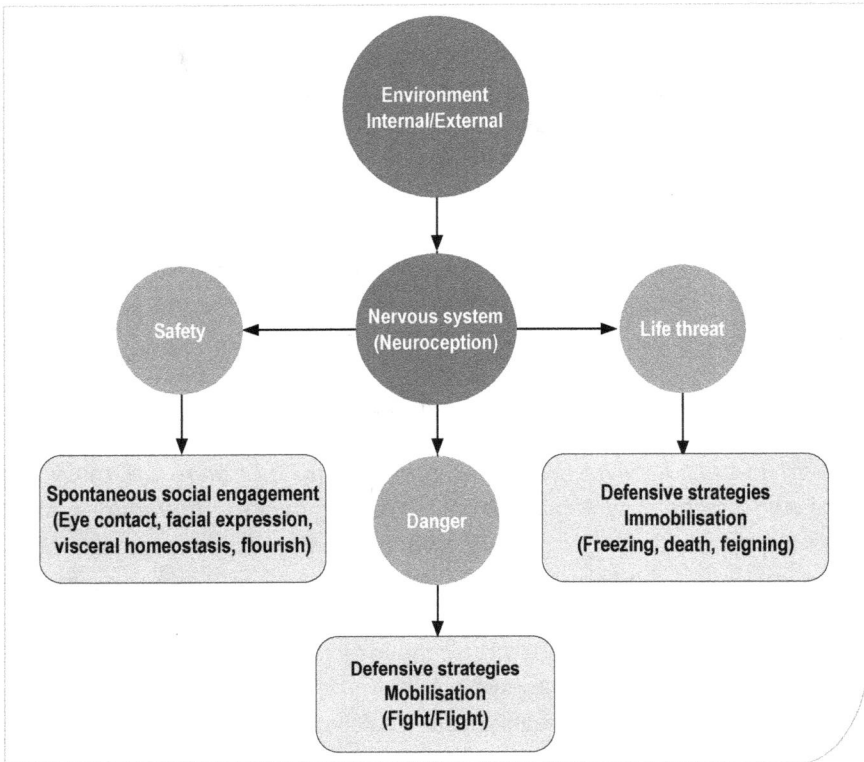

Figure 4: Three branches of the nervous system (Source: Adapted from Porges)[31]

The freeze response

The freeze response of hypoactivation is the oldest and most primitive instinct. You can see this response in reptiles. It is partially controlled by the oldest part of your brain – the brain stem, also called "the reptilian brain". In reaction to what is perceived to be a life threat, this branch activates immobilisation behaviours such as numbness, dissociation, fainting, feigning death and shutting down. This is also the most limited survival response in that it renders you incapable of actively defending yourself. In this state you are inactive, your metabolic rate is lowered, you have very low energy (hypo-aroused), and physical movement is dramatically reduced or non-existent. While this might work well for reptiles which are cold-blooded and can easily survive lowering their metabolic rate and playing dead, this does not work as well for humans!

The symptoms of the freeze/immobilisation response (a.k.a. hypo-arousal) are as follows:

- Shutdown/reduced awareness of sensation.

- Emotionally numb or flat/disconnected.

The fight-or-flight response

Next to evolve is the fight-or-flight stress response. This defensive response, which all mammals share, is an active defence strategy – you can fight or run away to save yourself.

In this state your heart rate increases, your digestion shuts down, your pupils enlarge and blood is directed toward your muscles; in short, your physiology is being primed to actively fight or flee. In this state, your energy is high (hyper-aroused) and you are ready for action.

The symptoms of the fight-or-flight response (a.k.a. hyper-arousal) are as follows:

- Fight or flight/emotionally overwhelmed/rigid and inflexible/impulsivity/tension.

- Anxiety/panic/emotional outbursts/overeating or restricted eating/obsessive rumination/rage.

The flourish or social safety response – from protection to connection

Finally, the most newly evolved branch, which only mammals share, is called the social engagement system or social communication system. This comes online when you and/or the environment feel safe. Whereas in the previous two survival responses your nervous system is shouting at you to fight, run, or pretend you are dead, this survival response system primes you for connection and communication with others. Being in this state enhances your ability to truly bond, attach and be emotionally close to others. In addition, your ability to be present and more at ease is equally enhanced, so you can enjoy life in real time.

We are wired to connect with others and to search for give-and-take relationships. This is where you move out of constant threat detection and threat response (for example, appeasing others) and into authentic engagement with others. According to Porges[32], connectedness is a biological imperative. Rather than your physiological arousal level being too low or too high, in the social engagement system your physiology swings into a more regulated and homeostatic state. Here you would feel a sense of calm. This state is also where you shift away from ruminating or catastrophising about problems, endless worry or stress. Being in connection with others enhances your ability to be regulated sufficiently so that you can relax and access a wide range of bounce back resources, rather than being a slave to your unceasing fight, flight or shutdown responses.

The Polyvagal Theory essentially demonstrates that pro-social behaviour, social communication, and visceral homeostasis (which is where you heal and repair your body) are not accessed in either of your defensive survival responses. These only occur when you are more regulated and have activated the social engagement system. From a polyvagal angle, the social engagement state is the optimal state in which to live your day-to-day life, to secure the many benefits from being connected to others in a caring give-and-take relationship, and to sustain a happier, nicer and more interpersonally attuned you.

Our daily experience is deeply influenced by whether or not our survival responses have been activated. By having a basic understanding of the evolution of your automatic survival responses, you are in a much better position to work with your nervous system rather than against it.

4.4 An electrochemical symphony: your enemy and your friend

Neuro-chemistry creates (and is structured by) our behaviour, and raises fascinating questions of whether each of us is in charge of our brain or our brain is in charge of us. From moment to moment during the day, the brain receives information via electrical impulses regarding the body's chemical balance, temperature, oxygen use, and so on. This mass of raw data activates chemical whirlwinds and streams of electrical discharges. These sparks run up and down the spidery webs of nerve cells. When an impulse arrives at the brain, the cortex organises the raw data, creating even more complex patterns or electrochemical impulses to make the data meaningful. The cortex does not inform us as to how this continuous data processing happens in its grey matter. We can get a glimpse of it through the kind of EEG (electroencephalograph) information used in neuro-feedback training.

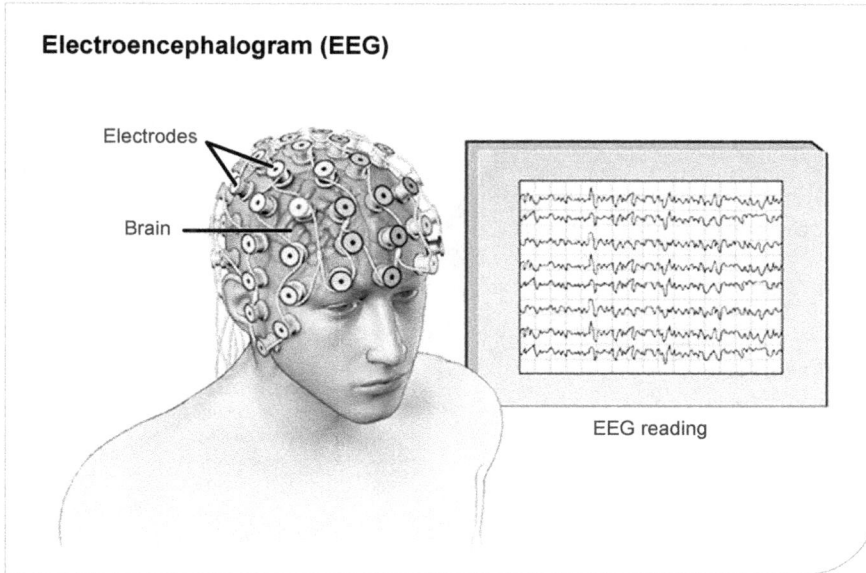

Figure 5: The mechanics of an EEG reading (Source: HVMN)[33]

Our fluctuating mental states are reflected in these electrical brain wave patterns. EEG readings provide a measure of brain activity which produces electrical fields that have a quantifiable number of cycles per second (frequencies) and equates to different types of paying attention. There are five common categories of brainwaves, and each is identified by the number of up/down wave cycles measured per second (Hertz/Hz). See Figure 6 for the different brainwave frequencies associated with different neurotransmitters or neurochemicals, which we experience as bodily sensations and emotions.

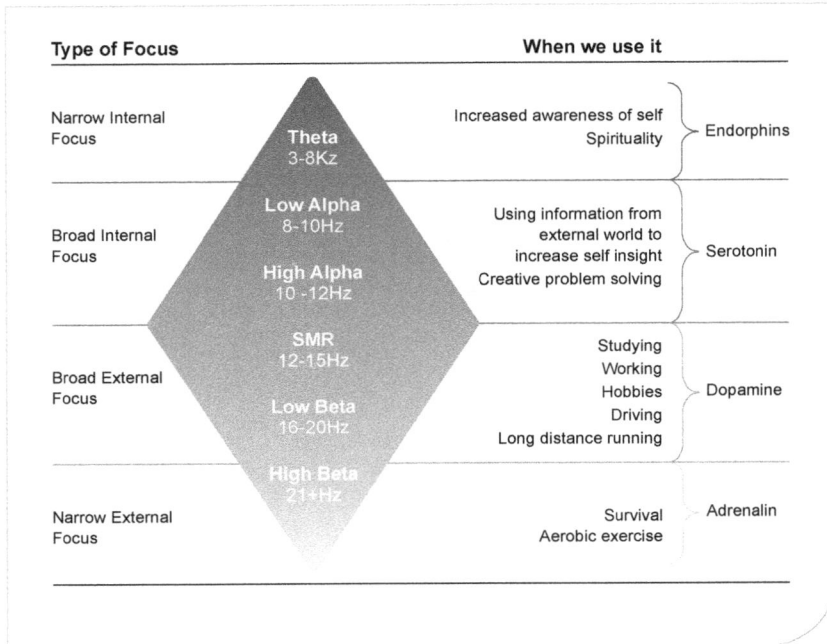

Type of Focus		When we use it	
Narrow Internal Focus	**Theta** 3-8Kz	Increased awareness of self Spirituality	Endorphins
Broad Internal Focus	**Low Alpha** 8-10Hz **High Alpha** 10-12Hz	Using information from external world to increase self insight Creative problem solving	Serotonin
Broad External Focus	**SMR** 12-15Hz **Low Beta** 16-20Hz	Studying Working Hobbies Driving Long distance running	Dopamine
Narrow External Focus	**High Beta** 21+Hz	Survival Aerobic exercise	Adrenalin

Figure 6: The focus diamond (Source: Adapted from Zhuang et al.)[34]

Neuro-chemistry

Each of the dominant mental states has associated neurotransmitters or neurochemicals which have specific functions; neurotransmitters are either excitatory or inhibitory. Neurotransmitters transfer signals between neurons, and they either stimulate or calm you. The excitatory neurotransmitters enable human drive and energy, while inhibitory neurotransmitters have the opposite effect – they keep you cool and calm under pressure.

Neurotransmitters use hormones to send messages to your organs, keeping your lungs breathing and your stomach digesting food. Neurotransmitters are like the gas for your car; they need the electrical spark from your nerve cells. Like your car's oil, hormones keep processes working smoothly. Your brain evolved to deal with a particular set of challenges, i.e. it controls your body's disparate systems, integrating them into a unified whole.[35]

There are currently over 50 substances that we know of that operate as transmitters in the brain. Drugs and medication can magnify, decrease or inhibit their influences. Table 1 provides a short list of transmitters and their actions. (To learn more about neurotransmitters, see Appendix B.)

Table 1: Selected neurotransmitters (Source: Ghadiri, Habermacher & Peters)[36]

Neurotransmitter	Impact/Function
Acetylcholine	Memory, attention
Serotonin	Fear, decision-making, mood
Noradrenaline	Energy, mood, attention
Dopamine	Feelings of reward, attention
Endorphin	Well-being
Oxytocin	Trust, love
Corticosteroids	Stress, anger
GABA-deficiency	Fear disorders
Testosterone	Dominance, aggression

The neuro-chemistry of our behaviour at work

Recall the last time you battled to concentrate in a meeting; the neuro-chemical footprint of struggling to concentrate can indicate a dopamine deficiency. Dopamine is one of the neurotransmitters synthesised by a small group of neurons in the brain and is the neurotransmitter linked to novelty seeking, desire and reward. EEGs show a spike in brain activity (13HZ- 16HZ) or (gamma – low beta) which is accompanied by the release of dopamine (in the frontal lobe) when there is a thought process of 'curiosity' taking place. However, since the brain is efficient and takes short cuts on what it already knows as soon as the new solution, insight or experience becomes familiar, it is 'categorised' by the brain as 'business as usual', resulting in a smaller electrochemical reaction. This means that the 'feel good' factor diminishes as you become used to the experience, approach, solution or situation.[37]

Therefore, to ensure that the brain keeps finding novel solutions and to be vigilant regarding 'categorisation', novel experiences need to be cultivated. This can be done by confronting the perceptual system with people, places and things it has not seen before – like asking a question which you do not have the answer for already.

However, to experience it as a 'feel-good encounter' where dopamine production is stimulated, the mind must be calm and relaxed. The moment one experiences a hyper-focused state with accompanying feelings of agitation and irritation, the dopamine is converted into adrenalin and this process cannot be reversed; the innovative thought will be turned in to narrow 'black and white' thinking.

You need to be aware that adrenaline causes memory to deteriorate. The best way to deal with this type of situation is to take a break and do some relaxing, such as diaphragmatic breathing, for at least five minutes. That will help achieve the calm, integrated state of focus needed for problem-solving.

To conclude, to try and motivate yourself through stress (e.g. adrenaline), you really need to think again. This is like shooting yourself in the foot while trying to run. The brain needs the slow electrical frequencies and the chemicals associated with relaxation to perform high-level intellectual functions, like finding innovative solutions.

When the brainwave frequency distribution is harmoniously balanced, one is calm, focused and 'in the zone'. This refers to a parasympathetic state in the central nervous system and optimal physical, intellectual, emotional and spiritual functioning – a 'flow state' or 'peak performance'. Instead of unnecessary and continuous adrenaline production, the body automatically starts producing the chemical substances which help one to experience physical well-being, energy, happiness, contentment and joy.

Conclusion

The "triune brain" theory describes the evolution of brain development, showing that there is a sequence to brain development. A deeper dive into the electro-chemistry of the brain helps to understand the safety-first principle. Our brain chemistry and our brain waves go together. Mental states (electrochemical activity) precede perspective, attitudes, beliefs, emotions and behaviours. Through this knowledge of the electrochemical brain, we can learn to change our experience of 'reality'.

5

The Big 5 Neuro-Insights

This chapter focuses on five foundational concepts or Neuro-Insights that impact how we understand individual, team and organisational behaviour, as set out in Figure 7 below.

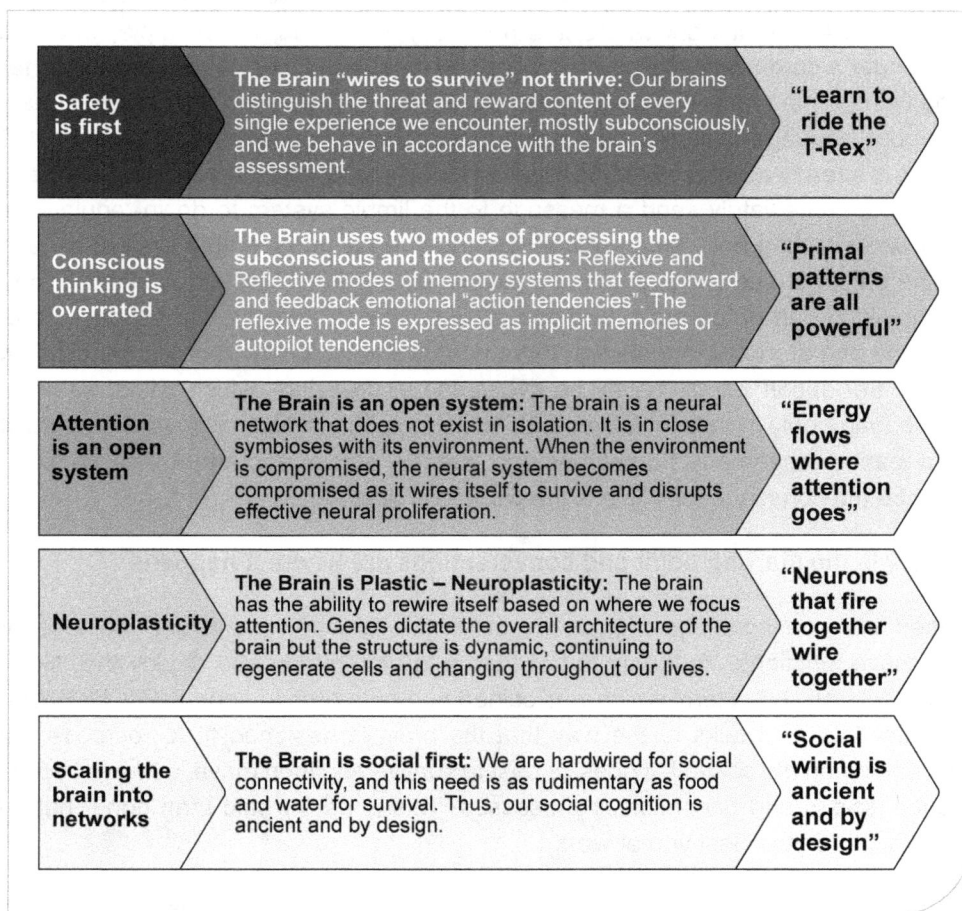

Safety is first	**The Brain "wires to survive" not thrive:** Our brains distinguish the threat and reward content of every single experience we encounter, mostly subconsciously, and we behave in accordance with the brain's assessment.	**"Learn to ride the T-Rex"**
Conscious thinking is overrated	**The Brain uses two modes of processing the subconscious and the conscious:** Reflexive and Reflective modes of memory systems that feedforward and feedback emotional "action tendencies". The reflexive mode is expressed as implicit memories or autopilot tendencies.	**"Primal patterns are all powerful"**
Attention is an open system	**The Brain is an open system:** The brain is a neural network that does not exist in isolation. It is in close symbioses with its environment. When the environment is compromised, the neural system becomes compromised as it wires itself to survive and disrupts effective neural proliferation.	**"Energy flows where attention goes"**
Neuroplasticity	**The Brain is Plastic – Neuroplasticity:** The brain has the ability to rewire itself based on where we focus attention. Genes dictate the overall architecture of the brain but the structure is dynamic, continuing to regenerate cells and changing throughout our lives.	**"Neurons that fire together wire together"**
Scaling the brain into networks	**The Brain is social first:** We are hardwired for social connectivity, and this need is as rudimentary as food and water for survival. Thus, our social cognition is ancient and by design.	**"Social wiring is ancient and by design"**

Figure 7: The SCANS foundational concepts

5.1 Safety first

The brain is wired for survival; the foundational organising principle of the brain is the motivation to 'minimise danger and maximise reward'. This motivation helps you deal with immediate threats, but also drives your search for rewards over longer time scales (from food to sex, relationships, purpose and meaning in life). It is important to note that it is mostly our primitive emotional circuitry that appraises all information for threat and meaning. This is a mostly subconscious process in which, in a split second, we label people, data or things as a friend/reward or a foe/threat.

Fear at work

We have an entire emotion dedicated to recognising and focusing on threats: fear. This inherent approach to seeking safety first results in our regularly experiencing fear in situations at work where it is not really warranted, for example, when you are under a chronically high cognitive load to meet deadlines. If your boss then tells you flat out that she is not there to pamper the team to perform, this added stressor (combat language) disrupts the smooth functioning of your Prefrontal Cortex (PFC), which is already activated but has a limited blood supply and therefore limited energy. It cannot immediately send a message to the limbic system to down-regulate (act cool while feeling hot). Much of the blood is already in the limbic system because of the earlier stress and is thus not available for the PFC. This stops blood going to the thinking part of the brain, and you struggle to think clearly. You fall back on gut feelings and autopilot tendencies. If the negative 'unsafe' environment persists, the only other option will ultimately be avoidance or 'freezing', which could mean that you engage only in safe initiatives with limited impact on both organisational growth and team engagement. Sometimes over-arousal occurs and aggressive behaviour may be the consequence. This is the worst you at work.

Safety is the starting point and conversations are where it happens

When our conversations trigger our instinctive or impulsive brain, we lose our executive functions. Your way out of fear is to understand and deploy the "Neural Axis of Creativity", a term which was coined by South African neurologist, Dr Etienne van der Walt.[38] It talks to the way that the brain is designed fit for purpose and, depending on the context, it does its best to survive and then thrive, using conscious, subconscious and unconscious processes. This is easier said than done, but it is possible to be the best you at work.

The bridge from foe or frenemy to friend in the corporate world is a trusting relationship which is underpinned by trusting and meaningful conversations (see Figure 8). We can prime our conversational partners for trust by tuning into others, setting good time boundaries, using resonant language, and behaving consistently.

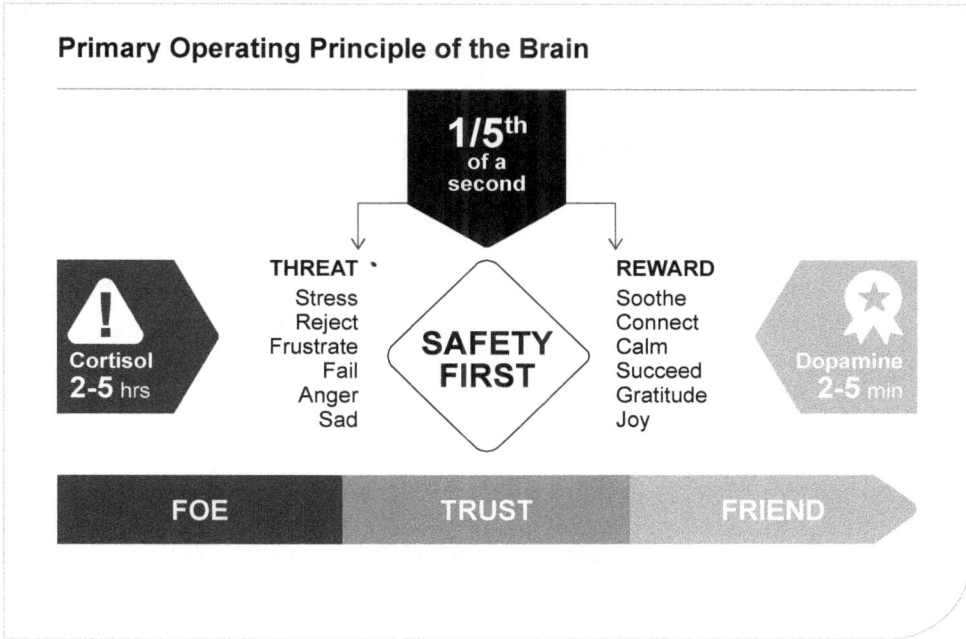

Primary Operating Principle of the Brain

1/5ᵗʰ of a second

THREAT		REWARD
Stress		Soothe
Reject	SAFETY	Connect
Frustrate	FIRST	Calm
Fail		Succeed
Anger		Gratitude
Sad		Joy

Cortisol **2-5** hrs

Dopamine **2-5** min

FOE · TRUST · FRIEND

Figure 8: The foe/frenemy vs. friend impulse

The friend/foe tagging has corresponding neurochemistry, for example, adrenaline and cortisol linger much longer in our neurochemistry than dopamine. You cannot manage yourself and should not even attempt at managing or leading others when you are overly adrenalised!

'Safety First' at work implies psychological safety. It is not an easy, non-demanding 'anything goes' culture, but it is epitomised by interpersonal risk-taking, where people feel able to speak up when needed – with relevant ideas, questions or concerns – without being shut down. Psychological safety is present when colleagues trust and mutually respect each other and feel able, even obligated, to be candid, while still adhering to high standards, meeting deadlines and learning. Neurochemically, it is filled with 'feel good' chemicals.

Neuro-hacks to build 'Safety First' behaviour at work

- **Know that safety is more than providing a physically safe environment at work.** Safety is also established by providing emotional support to peers, team members and the boss.

- **Become aware of your somatic markers**, such as your breathing rate, and consciously regulate that rate through slower breathing, which can automatically quieten the mind, resulting in the perceived experience of more safety.

- **Build relationships (respect and rapport) first**, then focus on the task at hand. When a new employee starts, ensure that there is a social welcome first prior to the security and admin controls and sign-ons.

- **Watch your tone of voice** – combat language and confrontational questioning can be threatening. A supportive, non-directive approach counters the discomfort attached to getting feedback by a person in power (i.e. the boss).

- **Cultivate psychological safety**, where colleagues trust and respect each other and feel able, even obligated, to be candid.

Conclusion

We humans are fundamentally in the business of pursuing pleasure and avoiding pain, so it is a fact – you need the negative to survive, but the positive to thrive. Despite the enormous complexity of our brains, the patterns of connectivity are highly ordered and stereotyped across individuals of the same species. We have developed neural patterns for survival, yet these patterns can be shaped into habits that represent safety – where we can be candid and caring, but still be high performing individuals.

5.2 Conscious thinking is overrated – the untapped power of your brain

"In the economy of action, effort is a cost, and the acquisition of skill is driven by the balance of benefits and costs. Laziness is built deep into our nature."

Daniel Kahneman[39]

The brain constitutes just 2% of body mass but taps into 20% of the body's energy resource through cognitive demands that draw more oxygenated blood into the brain, as neurons (the building blocks of the brain that are electrically active) need fuel to fire. To conserve energy consumption, the brain has developed ways to process information, such as the capacity to store learned skills in deep subconscious structures of the brain, where they can be automatically retrieved when needed in the form of habits, reflective routines, or cognitive biases.

The exceptionally low energy cost of this highly refined process enables the rest of the brain to optimise the use of oxygen and glucose to acquire new knowledge and solve complex problems creatively. The brain is also a master at calculating risk vs. reward ratios on a 24/7 basis and is a pattern-seeking device par excellence. These pattern-seeking behaviours have an application in understanding our behaviour at work as well.

The above quote from Kahneman relates to what he calls the machinery of the mind, which is in essence a dual processor of the brain, divided into two distinct systems that dictate how we think and make decisions. One is fast, intuitive, reactive and emotional, while the other is slow, considered, methodical and rational. While many choices can rely on fast thinking, occurring almost automatically and requiring little attention, like delegation of work, other decisions require slow, deliberative thinking, for example when merging with another company or rolling out a new strategic objective.

Different parts of the brain facilitate choices resulting from deliberate attention and those that are habitual. Slow thinking engages the prefrontal cortex areas, which are evolutionarily newer parts of the brain and involved in executive functions, whereas fast and habitual thinking relies on deeper, more primitive brain structures.[40]

Fast thinking also contains a non-conscious state – this level of self is shared by many species. This is the most basic level of awareness, which is signified by a collection of neural patterns that are representative of the body's internal state. The function of this 'self' is to constantly detect and record, moment by moment, the internal physical changes that affect the homeostasis of the organism.[41] This is also called cellular memory.

Fast decisions rely on fast thinking, which occurs almost automatically and requires little attention. To enable this fast thinking, the brain relies on stereotyping – placing people, places and objects into groupings such as 'familiar', 'foreign', 'in' and 'out'. The reflexive mode is expressed as implicit memories or 'autopilot' tendencies.

Predictably irrational

We are far less consistent, objective, fair and self-aware in how we navigate the world than we think we are.

This conceptualisation of the brain as a dual system that uses two modes of achieving optimal processing, the subconscious and the conscious, is helpful to understand behaviour at work for the following reasons:

- Much of our behaviour entails an interaction between the reflexive/implicit (system 1) and reflective/rational (system 2) modes.[42, 43, 44]

- The X-system (reflexive) gives 'feedforward' in relation to emotional 'action tendencies', i.e. we act on subjective cues within a spilt second.[45]

- On the other hand, the C-system (reflective) can exert control over the X-system – the mechanism underlying self-control processes like impulse control and emotional regulation.[46]

The reflexive (X) and reflective (C) thinking modes (see Table 2 below for a comparative summary) reinforce the idea that reflection occurs when our unconscious expectations are not met. In other words, reflective processes (the C-system) can be functionally demarcated as those designed for and recruited to handle situations that prove too difficult for reflexive processes (the X-System). Lieberman[47] also introduced the concept of quality in the reflective process, in that quality is based upon the cognitive load (the capacity of the brain and how that capacity is utilised) and level of motivation. Higher quality reflections are also more likely to be remembered. A way to integrate the X- and C-systems is to slow down and reflect on what factors are driving you in a particular direction, or to broaden your thinking by developing a habit for considering alternatives, asking questions, or asking others for input.

Table 2: Reflexive vs. reflective thinking modes (Source: Lieberman)[48]

Reflexive/FAST	Reflective/SLOW
Nonreflective consciousness	Reflective consciousness
Effortless	Effortful
Parallel processing	Serial processing
Gut feelings	Cognitive control
Spontaneous processes	Intentional processes

Trusting your gut

Gut feelings are messages from mostly the insula and the amygdala, which the neuroscientist Antonio Damasio called, "somatic markers".[49] These markers are feelings in the body that strongly influence subsequent decision-making. The somatic marker hypothesis proposes that emotional processes guide (or bias) behaviour, particularly decision-making.

These messages are sensations that something 'feels' right, such as the association of a rapid heartbeat with anxiety or nausea with disgust. Somatic markers simplify decision-making by guiding our attention toward better-known options, but they are hardly fail-safe.

We are memory systems – for better or for worse

> "Memory has always fascinated me.
> Think of it. You can recall at will your first day in high
> school, your first date, and your first love."
>
> Eric Kandel[50]

In his book, In search of memory, Eric Kandel (a pioneer of investigations into the biological nature of memory, and a Nobel Prize winner in 2000 for his contributions to neurobiology) described memory as, "a form of mental time travel [which] frees us from the constraints of time and space".[51] It is the brain's amazing ability to store a seemingly infinite number of facts, figures and experiences which enables us to travel back in time and across space to retrieve the autobiographical material which defines what we pay attention to or not. Kandel's work on memory systems has revolutionised psychotherapy and has influenced the field of leadership development and organisational behaviour.

How we act is a function of what we see, which, in turn, is a function of deeply-held memory systems. These memory systems are unique, and are socially constructed and pattern-seeking. We are predisposed to be negative towards what is deemed different, and therefore become quite attached to our own mental short cuts (implicit memory systems/reflexive reasoning/baby blueprints).

We also make attributional errors which prevent us from having an objective view of reality. Facing reality as it is, versus how we want it to be, requires being an observer to our own thoughts and actions. It requires making the unconscious conscious so that we can look at it and work through the undigested emotional baggage to regain a part of ourselves. This can be done through coaching, journaling and mindfulness practices.

#Neuro-hacks for leveraging conscious and unconscious thinking

- Everything you think, say and do is a result of **instantaneous processing**. Know that optimal human functioning implies an ongoing combination of unconscious and conscious processing that underpins all your brain's key circuits.

- Develop **a reflective practice** to digest your daily, moment-to-moment habits. Knowing and training your brain to be able to work more effectively with these processes means living with an awareness of how to be more of your best self and less of your worst self at work.

- **Spot it you got it = perception is projection**. The key to self-knowledge and awareness of your own unconscious drivers is to be aware that what you are trying to fix in others is what is wrong in yourself... and what you admire in others, you also have.

- **Consciously confront** your brain's reliance on categories/biases – write down your gut feelings and know that they are not fail-safe.

Conclusion

To leverage the subconscious mind and all its superpowers, you need to still your mind out of the grip of the unconscious primitive brain and unprocessed emotional baggage. The more stressed you are, the more the control in the brain sinks to the limbic system and the brain stem (the instinctive, impulsive brain). The more relaxed you are, the more the conscious rational cortex (intuitive, clever brain) can control your behaviour. Relaxation activates the slower brain waves associated with calmness, insight, wisdom, larger perspective, and creativity.

5.3 Attention is an open system

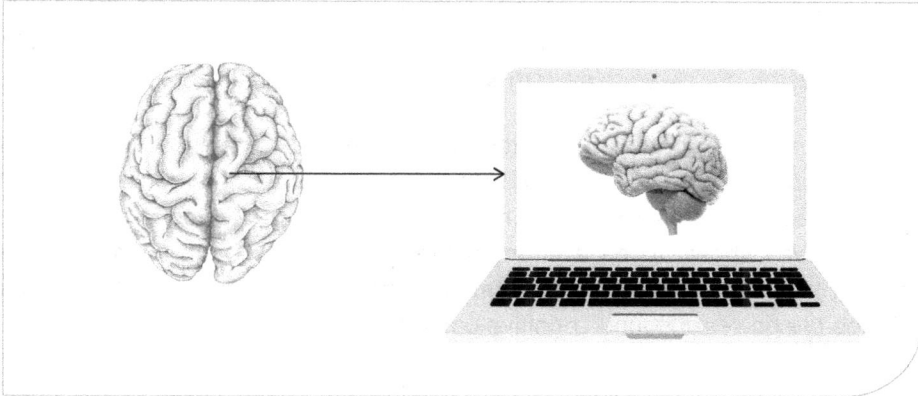

"Attention is like a combination spotlight and vacuum cleaner: it illuminates what it rests upon and then sucks it into your brain – and yourself."

Rick Hanson[52]

The skill that stands at the top of human evolution

Attention might seem like a dull subject, but it is the chief mental tool through which we direct our awareness and experience the world. The term "paying attention" is very appropriate for it is often more costly than we realise. Not paying attention appropriately can cost us in the form of emotional problems and physical ailments, and may prevent us from reaching our full potential and flourishing in life.

What strictly sets us apart from all other forms of life is the human ability to pay attention to how we attend. This skill stands at the top of human evolution. Attention skills let us alter our reality; we are the architects of our lives in a very real sense. One minute we can be an inclusive, loving being, and the next an adversarial bully filled with combat language. What accounts for these changes? It is simply a spontaneous change in our style of attention. We are affected by shifts in attention all the time, but we usually do not realise that changes in how we feel have to do with changes in the way we attend to thoughts, emotions and physical sensations. These shifts in attention – in the way we shape and direct our awareness – play a huge and unrecognised role in our lives.[53]

Attention is a mental muscle; like any other muscle, it can be reinforced by the right kind of exercise. Because of neuroplasticity, whatever you hold in your focus of attention physically changes your brain. Seeing that we have a negativity bias for survival, attention sucks up the negative first, resulting in us becoming supercharged

31

on the negative. Attention is also a finite resource and it can easily be hijacked by emotional reactions. However, thought is grounded in recent layers of the brain, so attention is under our conscious control; we can access it if we can educate and train the brain. We can learn how to change the way we pay attention and this can have robust effects on our nervous system which impacts our quality of life.

A 15-year longitudinal study conducted in Dunedin, New Zealand, on the cognitive control of more than 1,000 children, with a follow up when they had reached their 30s, revealed surprising results.[54] These children's ability to focus in the face of distraction was shown to be a stronger predictor of their adult financial success than both their IQ and the financial status of their families, a finding that strikingly illustrates the power of sound cognitive control.

When we develop our attention skills, we can lay down our troubles at any time and leave them. Or, if we want, we can take them up again. If we are anxious, depressed or experience chronic physical pain, we can allow it to diffuse by changing the way we attend to it. Ultimately, by learning to use a variety of attention styles (I will focus on the two main ones), we can free ourselves from the guilt and shame of the past and stop being afraid of the future.

"Energy flows where attention goes"

Open focus attention

Learning to foster attention that is non-exclusive and non-judgmental supports the integration of your experiences with openness and flexibility. This takes practice. An open and relaxed focus also encourages creativity and innovation since a single and sharp focus on goals will not necessarily yield the expected and desirable results. You will not move from 'DUH' to 'AHA' if you cannot induce choiceless awareness. So, self-awareness is crucial and allows us to monitor whether our mode of attention is suitable for the situation at hand.

Addiction to narrow focus

In today's supercharged world, we live with technology that enables us to deal with an almost unending stream of seemingly urgent messages and data relating to emails, appointments and a depth of decisions, all while trying to be nice to others.

These numerous distractions draw attention away from more immediate and crucial tasks and demands. Maintaining a clear focus on these tasks and demands requires that the brain's circuitry be consistently engaged and occupied for attention. To harness the necessary mental attention needed to process our cognitive load or information load, cognitive effort is required. The urban dictionary defines cognitive load as: "1. A state of extreme mental exhaustion because of excessive content absorption or mind-numbing work or 2. the state of your current brain power for any particular task or event."[55] For example, you can struggle with a serious cognitive load when you are a sunshine yellow person and you have to reconcile cool blue financial accounts.

Attention, as with body muscles, can, however, also become fatigued, which can result in symptoms such as lowered effectiveness, increased distractedness and irritability – symptoms that also indicate depletion in the energy levels needed to sustain sound neural functioning. To select a single point of focus and resist the attraction of all else, however, calls for "concentration", which requires that we be able to sift through irrelevant information to decide what is important. Those who have mastered this art are able to remain energised by avoiding distraction and resultant attention fatigue.

Yet a chronic, narrowed focus of attention creates a behavioural loop.[56] Narrow focus worsens fearful conditions, and then when circumstances change and we are no longer in 'danger', we tend to stay in narrow focus as a way of circumventing our residual fears and anxieties. These residual fears are typically accompanied by a middle- to high-beta range of frequencies to keep unpleasant feelings from surfacing.[57] Thus, narrow focus is used as a tactic to escape. As feelings of anxiety increase we unconsciously look for effective distractions to keep us from feeling them. We pin attention on our emails, social media or Netflix series in part to escape emotional chaos, anxiousness, or unpleasantness from within. The more interested we are in something 'out there', the more effective it is as an anxiety-management technique.

Multi-tasking equals multi-failing

"Any man who can drive safely while kissing a pretty girl is simply not giving the kiss the attention it deserves."

Albert Einstein[58]

Does this sound familiar? You are always busy at work. You never have adequate time to finish what needs to be done and are always doing three things at once. You make mistakes and have to redo a lot of your work. What is going on? Different brain functions allow us to either focus our attention exclusively on one task (focused attention) or divide it between multiple tasks. Since human attention is limited, trying to divide it between too many tasks necessarily results in performance errors.

It is biologically impossible to learn, study or recall information to which the brain has not paid attention. The brain is not designed to remain attentive and focused on the same stimuli for extended periods of time. Such circumstances yield diminishing returns over time. The brain will interject its own downtime in order to strengthen new synaptic connections related to attention. Downtime is not the same as multi-tasking.

Multi-tasking is an illusion (a misnomer) in that neurons do not fire properly and then do not wire. The reality is that we are switching back and forth between tasks (Worringer et al., 2019).[59] Switching happens so fast that it appears we are performing multiple tasks simultaneously – like the concurrent performance of several jobs by a computer. We need both recognition and recall memory to have clarity and certainty of mind; when you multi-task, both recognition and recall memory are compromised. Multi-tasking thus equates to multi-failing. Workers who are multi-tasking by paying continuous partial attention to emails and phone calls suffer a fall in IQ more than twice that found in marijuana smokers.

Pay attention to how you pay attention to others

The word 'attention' originates from the Latin *attendere*, which means 'to reach toward'. This is a beautiful definition of focus on others, which is the foundation of empathy and of an ability to build social relatedness. The brain is a neural network that does not exist in isolation; it is in close symbiosis with its environment. When the environment is compromised, the neural system becomes compromised as it wires itself to survive and disrupts effective neural proliferation.

This is also referred to as the field of epigenetics (epi = that which sits above (genes)). Both genetics (template genes) and the environment (transcription genes) interact in the brain to shape our brains and influence our behaviour. Our genes (nature) are not our destiny – they are our disposition templates and you ignore them at your peril. Your bonding and conditioning experiences ('nurture') can have a lasting effect from childhood that shapes your brain's ongoing personal experiences throughout your life, but importantly, your brain has a remarkable capacity for change – 'plasticity'. With the right insights and training, transformative brain change that translates into new behaviours is possible. We now know that we are hardwired for social connectivity, and this need is as rudimentary as food and water for our survival. Thus, our social cognition is ancient and by design. Our social relationships (also at work) modify our neural systems and enhance or distract our attention. So, pay attention to how you attend to others – especially your nearest and dearest.

#Neuro-hacks for attention deployment at work

- Pay attention to the way you pay attention:

  - Be aware of when narrowed/focused attention is your default. Focusing on one or a few things as the foreground and making everything else the background sounds good as it allows us to perform some tasks very well, but it can very quickly become an addiction with negative consequences for the nervous system. Not everything is urgent and critically important, but we can treat it that way. The price is a frazzled nervous system and a volatile life.

  - Cultivate open focus attention or choiceless awareness, which helps to develop your attentional skills. When we practice and deploy open focus attention we develop an awareness of how we attend to a wide array of sensory experiences, as well as the spaces between those experiences.

  - Develop your controlled attention, or the capacity to stop automatic reactions and thoughts as needed, through mindfulness practices. Mindfulness improves three qualities of attention – stability, control and efficiency. The human mind is estimated to wander for roughly half of our waking hours, but mindfulness can stabilise attention in the present. Individuals who complete mindfulness practice trainings are shown to remain vigilant longer on both visual and listening tasks.

  - Develop your sustained in order to resist distractions, for example, turn off your cell phone in meetings, limit meetings to 30 minutes where possible, ensure that there are adequate breaks during longer meetings, and set more frequent/shorter meetings. Consider having 1:1 meetings outside your own office to minimise distractions.

- **Stop multi-tasking** – it causes you to do multiple things poorly as you cannot focus on two things at once. *#multi-tasking is multi-failing*. PUT DOWN YOUR PHONE!

- Switch from continuous partial attention to sustained **present moment awareness**.

- **Encourage serial processing** through deep listening and incisive questioning. Parallel processing burns up mental bandwidth.

> *"Information consumes the attention of its recipients."*
>
> Herbert Simon[60]

Conclusion

Neural networks will be formed for everything to which we pay attention and nothing to which we do not. By actively deciding what warrants our attention, we are applying so-called 'top-down' attention. This is different to 'bottom-up' attention, which means that we are allowing our focus to be mechanically dictated by whatever appeals to it. Attention requires both narrow and open focused concentration, and is a prerequisite for neurons to be activated and neural networks to be forged. Forging new networks is energy-intensive, and our brains are not designed to remain attentive for long periods of time.

5.4 Lifelong neuroplasticity – the brain remodels all the time

"It is not the strongest of the species that survive, nor the most intelligent, but the one most responsive to change."

Charles Darwin[61]

The brain has the ability to rewire itself based on where we focus attention. Genes dictate the overall architecture of the brain, but the structure is dynamic, continuing to regenerate cells and make changes throughout our lives. The brain can change at any age – for good or for bad.

When we exercise our brains, we put our neurons into action. "'Cells that fire together wire together', meaning that synapses, or unions between neurons, become more solidified the more often the respective neurons 'talk' to each other. This is known as neuroplasticity enabled by the Hebbs rule, proposed in 1949 by the psychologist Donald Hebb[62] as a theoretical mechanism for how neuronal circuits are modified by experience."

Neuroplasticity is a process, not a single measurable event. If you do something once, a loose group of neurons will form a network in response, but if you do not repeat the behaviour, it will not 'carve a track' in the brain. When something is practiced repeatedly, those nerve cells develop a stronger and stronger connection. Like forging a path through tall grass by walking it again and again, it gets easier and easier to fire that network. This can be advantageous – it's called learning – but it also can make it difficult to change an unwanted behaviour pattern.

Luckily, there's a flip side: nerve cells that do not fire together, no longer wire together; they lose their long-term relationship. Every time we disturb the habitual mental or physical process reflected in a neural network, the nerve cells and groups of cells that connected to each other start breaking down their relationship.

This is like an experience most of us have had. When you resign from a company, you part with colleagues (or even your best friend at work) with whom you've shared so much, and you promise to stay in touch to maintain your connection. As time passes, you start sending them birthday wishes only, and the relationship begins to weaken and fade. This effect is an exact reflection of what is going on inside the brain. As you think less and less about the ex-colleague, the neural connections lessen, until there's no connection at all. What's happened is that the very fine dendrites spreading out from the cell body that connect to other cells unhook and are available to rehook to other nerve cells, letting the old patterns go and potentially forming new ones.

Neuroplasticity offers us a real opportunity to develop and grow new ways of thinking and acting. By becoming neurally aware through learning how the brain works, what affects the reward and action systems in the brain, and the way that stress and other distractions, distortions and misunderstandings lead to sub-optimal performance, we can effectively change how we perform and engage at work.

Habits – Neural pathways

New habits or neural pathways are delicate and a relapse into default habits or pathways occurs easily. The challenge is to facilitate enough activation towards new patterns of firing for the default patterns to shift. The myelin sheath that covers your neural pathways gets thicker and stronger the more it is used (think of the plastic protective covering on wires); the more a pathway is used, the stronger the myelin and the faster the neural pathway.

How the leopard changed its spots

It is no easy matter to change deeply ingrained patterns of behaviour that have been practiced since childhood, especially if you have identified yourself with your personality and built up a well-developed system of defence mechanisms to guard it, e.g. rationalise, deny, project, attack, withdraw, procrastinate, humour, etc. Not easy, but possible. How? You must be able to see your personality as a collection of memory systems or learned behaviours based on mostly childhood conclusions and survival strategies. You must make a sincere decision to change your perspective and you must let down your defences and fully relax to open the doors of the old brain to new information.

The key guidepost to changing the brain and our habits lies in this quote by Eric Kandel, cited in Kandel et al.:[63]

> *"Down regulating distress and facilitating enriched environments enhances neural proliferation."*

'Distress' means the brain is constantly experiencing fear, worry and negativity (a compromised environment). If this is chronic (always present) it becomes a steady-state (homeostasis) in which survival becomes paramount and all other neural functioning suffers.[64] In this state of affairs, the brain does not exert effort on creating new neurons, as in a compromised environment the electrochemical ecosystem becomes toxic, causing neurological, physical, emotional, intellectual and spiritual degeneration (breakdown), which leads to suffering and more stress. With discipline and repetition, we can change our stressed steady-state to a "new steady-state".

We now know that social experiential factors shape the neural circuits that are essential to social and emotional behaviour from the prenatal period to the end of human life. These behavioural experiences include both incidental influences, such as early adversity, and intentional influences that can be produced through specific interventions like mindfulness, talking therapies like coaching, and social inclusion – these influences promote enriched environments and well-being.

By facilitating controllable incongruence (where stress can be contained and the intuitive clever brain is activated) through keeping yourself and your team, children, and partner in a state of support and challenge, you are unlocking the key to neurogenesis – new neurons, pathways, habits, growth and expansion.[65]

Experience-dependent neuroplasticity

Key principles of experience-dependent neuroplasticity have been developed especially for rehabilitation after brain injury (see Table 3 below). These principles are also useful for learning in the intact brain.[66] Neuroplasticity is believed to be the basis for both learning in the intact brain and relearning in the damaged brain that occurs through physical rehabilitation. Neuroscience research has made significant advances in understanding experience-dependent neuroplasticity, and these findings are beginning to be integrated with research on the degenerative and regenerative effects of brain damage. The qualities and constraints of experience-dependent neuroplasticity are likely to be of significant relevance to rehabilitation efforts in humans with brain damage. However, some research topics need much more attention in order to enhance the translation of this area of neuroscience to workplace application and practices.

Table 3: *Major principles of neuroplasticity* (Source: Klein & Jones)[67]

1.	Use it or lose it	Failure to drive specific brain functions can lead to loss of abilities.
2.	Use it and improve it	Training that drives a specific brain function can lead to improving abilities.
3.	Specificity	The nature of the training experience dictates the nature of the change in the brain (*plasticity*).
4.	Repetition matters	Change (*plasticity*) requires sufficient repetition for lasting neural change.
5.	Intensity matters	Change (*plasticity*) requires intensive training.
6.	Time matters	Different forms of change (*plasticity*) in the brain happen at different times during training.
7.	Salience matters	The training experience must be meaningful to the person in order to cause change (*plasticity*)

39

8.	Age matters	Training-induced change (*plasticity*) occurs more readily in younger brains.
9.	Transference	Change in function as a result of one training experience can even lead to learning other similar skills.
10.	Interference	Brain changes (*plasticity*) that result in bad habits can interfere with learning good habits.

Doidge[68] provided evidence that reflecting and narrating to another person fosters learning and development, because of the significant change effect that language and meaningful social relationships have on the brain's form and functions (enhancing neuroplasticity). The principles of neuroplasticity thus come to life with and through enriched conversations with others. Coaching, mentoring and robust conversations at work are all ways in which neural pathways can be awakened and strengthened.

Neuroplasticity and memory

When we pay attention, the brain changes as memories are formed, stored and recalled in a complicated process that engages multiple circuits of the brain. Without memory, there is no learning. From the point of view of neuroscience, learning is a physical experience which involves changing the brain; the brain learns and builds new knowledge by forming memories.

Memory can be described as a three-step process of encoding, storage and retrieval. Incoming data are held in short-term, or working, memory, and are quickly lost if not fused with current information (called consolidation). How well we encode a memory is critical for our capability to recall it at a future point. Failure to learn can be a function of flaws at any of the three stages in the memory process.

#Neuro-hacks for lifelong neuroplasticity

- **The fail-safe recipe for neuroplasticity** – Focus on what you really want and then follow up with actions to entrench the new wiring. Be inspired and then perspire to make it happen. *#The basic math: Repetition x Intensity = Results.*

- Utilise **"chunking"** when attempting to learn complex information. Make use of different modalities when learning (visual, tactile, and auditory).

- **Embrace healthy-balanced living** – this leads to neurological, physical, emotional, intellectual and spiritual regeneration (growth).

- **Move it!** As counterintuitive as it seems, sitting does not conserve energy; it robs us of it, making us feel more tired. Short bursts of movement, on the other hand, produce a powerful energising effect by "tricking" the body into firing up its natural energy again. This creates the conditions for neurogenesis.

- **Sweat!** Oxygen alone is a vital resource for the brain, but when you do regular exercise that you enjoy, you release a neuroprotective growth factor called brain derived neurotrophic factor (BDNF). This is like fertiliser for the brain and is responsible for brain cell repair, regulating your mood, learning and memory as well as neurogenesis.

- **Stop complaining** – MRI scans have shown that constant complaining can lead to the shrinkage of the hippocampus – the region in your brain responsible for cognitive functioning. The smaller your hippocampus, the more likely you are to have your memory decline, as well as difficulty adapting to new situations.

- **Enhance your working memory** (the capacity to hold information "online" in the moment). An average adult is thought to have a memory span of holding seven digits (plus or minus two) items "online". One of the best studied methods for improving verbal memory is through the use of 'chunking' strategies, in which items are recoded into meaningful units or 'chunks', e.g. 2 1 1 2 6 6 (six items) is easier to retain when remembered as 21 12 66 (three items).

- **Develop your recall memory** (the capacity to remember information in the short-term). For example, verbally repeat data once or twice, e.g. a person's name twice after you are introduced. Always try to relate new ideas/people to those that already exist in your world (e.g. linking a new face to someone you already know). This will allow other memory areas to "help out" with the recognition.

Conclusion

You can teach an old dog new tricks. The growing understanding of the nature of brain plasticity shows that we can learn new habits, however neuroplasticity does not happen in a vacuum. It is experience-dependent, and some experiences make a more significant difference than others.

5.5 Scaling the brain into networks

The body-mind is an integrated network. We used to think of the brain and nervous system as a hierarchical structure with a control centre (the brain) at the top, and a downwards spiraling cascade of lower structures, each having a less important function than the one above. We now tend to think of the body-mind as a network, where every system within the larger system has equal power to control the functioning of the entire network.

In a network, you can theoretically press any button to cause a fundamental change in the entire network. It is like a spider's web where touching any part will affect the whole web; every point is connected to every other point, and all points are equal regarding the potential to control the flow of information throughout the network. Every sub-system is a node or point of entry into the whole system, which is why there are so many different kinds of developmental routes, each having the potential to make a difference.

Large-scale brain networks are a collection of interconnected brain areas that interact to perform restricted functions.[69] Certain networks act as controllers or task switchers that coordinate, direct and synchronise the participation of other brain networks. On the other hand, other brain networks enable the flow of sensory or motor information and participate in the conscious execution of tasks. There are various proposed networks that work within the "default network" supports introspection and creativity; pleasure and delight triggers the "reward network"; emotion and habits shape the "affect network" and support intuition; and the "control network" enables self-control.

Bressler and Menon provided a synopsis of the three large-scale networks (see Figure 9 below):[70]

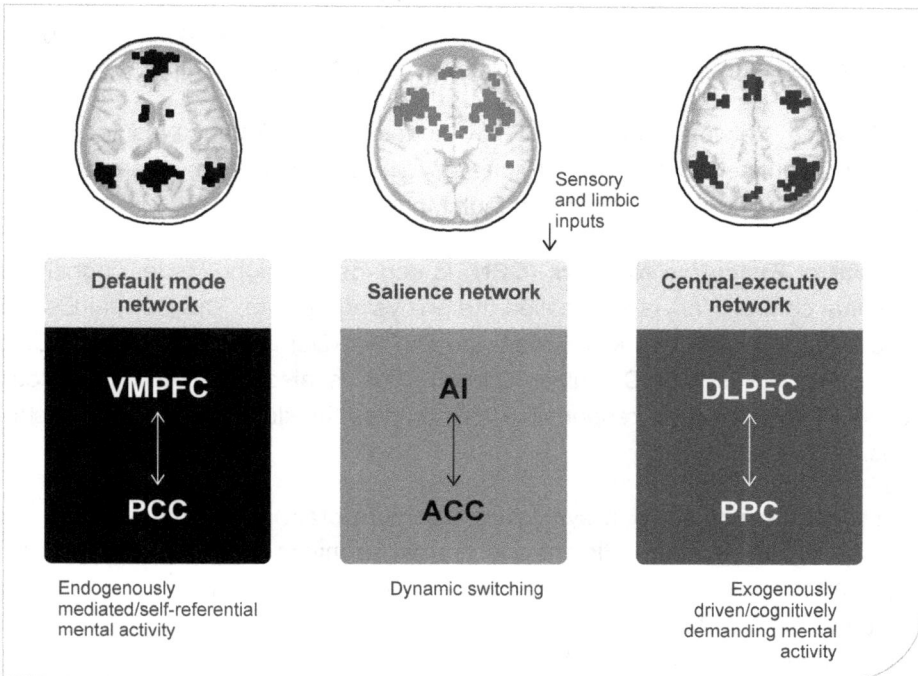

Figure 9: The brain as a network (Source: Bressler & Menon, photo courtesy of Steven Bressler)[71]

The Default Mode Network (DMN) is what the brain does when it is not engaged in specific tasks. The DMN comprises an integrated system for autobiographical, self-monitoring, and social cognitive functions. This is also called the mentalising system of the brain. The DMN is further responsible for rapid episodic spontaneous thinking (REST), which forms part of mind-wandering. Flying under the brain's conscious radar, the DMN regresses into the past and projects into the future, pays attention to the present tense, and recombines thoughts and ideas. Using this process is how we develop self-awareness and identity. It is the source of creative solutions and future predictions. The DMN also helps us tune in to other people's thinking, thereby cultivating team understanding and cohesion.

The DMN plays a pivotal role in emotional self-awareness, social cognition and ethical decision-making. The name 'Default Network' implies that we are social first and foremost. This social adaptation is central to making us the most successful species on earth. The Default Mode Network (DMN) has as key nodes the VMPFC (Ventromedial Prefrontal Cortex) and the PCC (Posterior Cingulate Cortex). This network (mostly in the middle of the brain) supports social thinking, such as self-awareness, collaboration, communication, trust and simulations, i.e. how we interpret the ways others are thinking, feeling and acting.

The Salience Network (SN) is a controller or network switcher. The controller decides which information is most important and which should receive priority in the queue of brain signals waiting to be sent, based on the task at hand.

The Salience Network (SN) has as key nodes the AI (Anterior Insula) and ACC (Anterior Cingulate Cortex). The SN initiates dynamic switching between the Central Executive Network (CEN) and the Default Mode Network (DMN).

The Central Executive Network (CEN) is engaged in higher order cognitive and attentional control. This is also called the analysing system. The Central Executive Network (CENW) has two key nodes: the DLPFC (Dorsolateral Prefrontal Cortex) and the PPC (Posterior Cingulate Cortex). This is also called the Task-Positive Network (TPN), which is responsible for complex thinking, learning and technical issues.

When these networks are in synchrony, optimum brain performance is the result. When synchrony is poor, efficient and normal cognition and motor behaviour are compromised. Neural activity in the CEN, is inclined to inhibit activity in the DMN, and vice versa. The TPN is triggered during a broad range of non-social tasks[72], and is correlated with focusing attention, making decisions and problem-solving.

Opposing domains

Being constantly goal-focused by using the analysing system switches off the brain circuits for thinking in terms of people – the mentalising system – which, ironically, are precisely the circuits that we need to have activated for managing and engaging with ourselves and others (see Figure 10). This is also called the opposing domains' neural seesaw[73]; these different networks have important repercussions for how we act, especially the way we lead others.

Another way of thinking about this is what may be referred to as exploitation thinking vs. exploration thinking. Exploitation requires concentration on the job at hand, whereas exploration demands open awareness to recognise new possibilities.

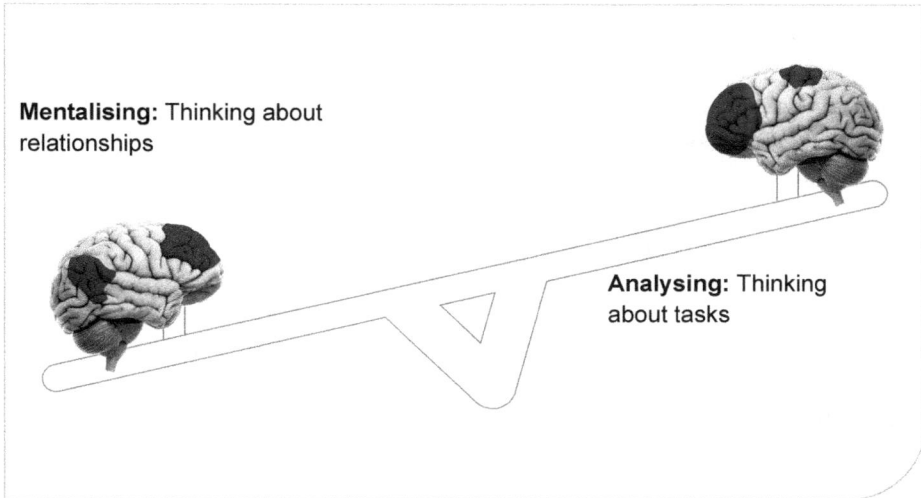

Figure 10: The neural seesaw (Courtesy of Metthew Lieberman)[74]

An article by Professor Richard Boyatzis, Professor of Organizational Behavior, Psychology, and Cognitive Science at Case Western Reserve University, and colleagues, entitled, *Antagonistic neural networks underlying differentiated leadership roles*[75], explored the link between the opposing domains theory and the long line of work that has been documented in the leadership literature since the 1950s (that of the task-oriented leadership role and the relationship-oriented leadership role). In their article, the authors raised questions about how we can effectively fulfill both roles. This anti-correlation between the TPN and DMN creates a fundamental neural constraint on cognition that is highly relevant to the different roles and capabilities that are required of us at work.

To be effective, we need to be aware of which network we are deploying, i.e. we must learn to apply Neural Resource Efficiency. Although not easy, becoming more mindful of these opposing domains and being deliberate in leveraging the strengths of each network for the task at hand is a turning-point in applying opposing domains theory.[76] A practical example is giving feedback which is a frequent activity at work; a brain-friendly way to do this cue is to keep task feedback separate from developmental feedback.

#Neuro-hacks for improving your balance on the neural seesaw

- **Do 'neural disposition' matching**. Identify what people's strengths are and match each person to the task at hand. Some people prefer to work with people, while others prefer to work with data and things.

- **Social skills are a great multiplier**. When you have strong social skills, you can leverage the analytical abilities of team members far more efficiently.

- **Co-activation is not easy, but is possible**. High-creative thinkers co-activate activity among the brain's executive, salience and default mode networks, showing stronger connections across both hemispheres of the brain. Creative thought is a whole brain network effort; it's the synchrony between these systems that seems to be important for creativity. People who think more flexibly and come up with more creative ideas are better able to engage these networks that do not typically work together and bring these systems online.[77]

Conclusion

Understanding large-scale brain networks, defined as a collection of interconnected brain areas that interact to perform circumscribed functions, can help explain behaviour at work. Learning to differentiate between task-positive and task-negative (socio-emotional) roles at work, as well as paying attention to the relevance of the network for the assignment at hand, can ensure optimum brain resource efficiency.

6

From scanner to the office

An explanatory model

Recent advances in neuroscience provide insights into strategies that can enhance well-being and performance, thus enabling us to shift from fight/flight/freeze behaviour to that of flourishing. The current organisational neuroscience, as well as social and affective neuroscience literature, was reviewed to confirm measurement scales currently used in these fields. A validation and regression analysis study was then done (N=208) using self-report brain-based diagnostics. The study found significant positive inter-correlations between scales related to what may be referred to as 'NeuroCapital' (as operationalised by seven brain-based self-report inventories). Significant negative inter-correlations were found between perceived stress and the scales related to NeuroCapital. See www.neurocapital.co.

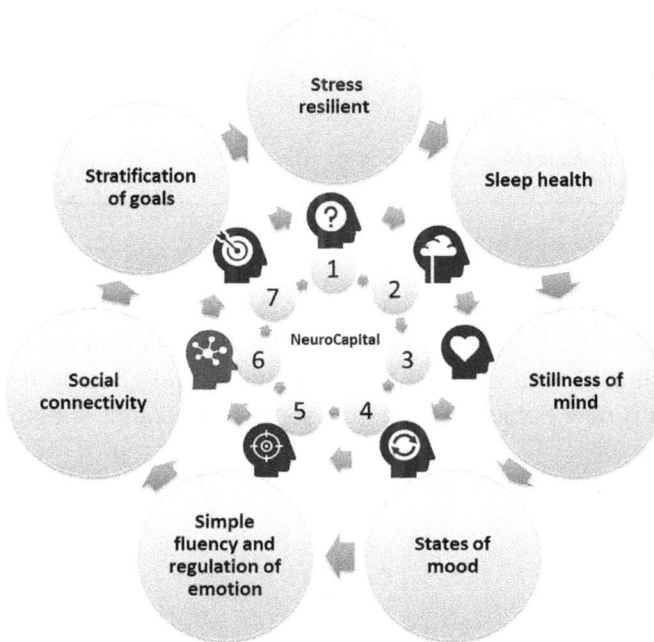

Figure 11: Seven strategies that cultivate NeuroCapital

These seven (surprising) strategies have differential and beneficial effects on the brain that complement each other, together providing a shift from fight/flight to flourish.

NeuroCapital can be measured and developed for individuals, groups and organisations. A conceptual model is provided in Figure 12 below:

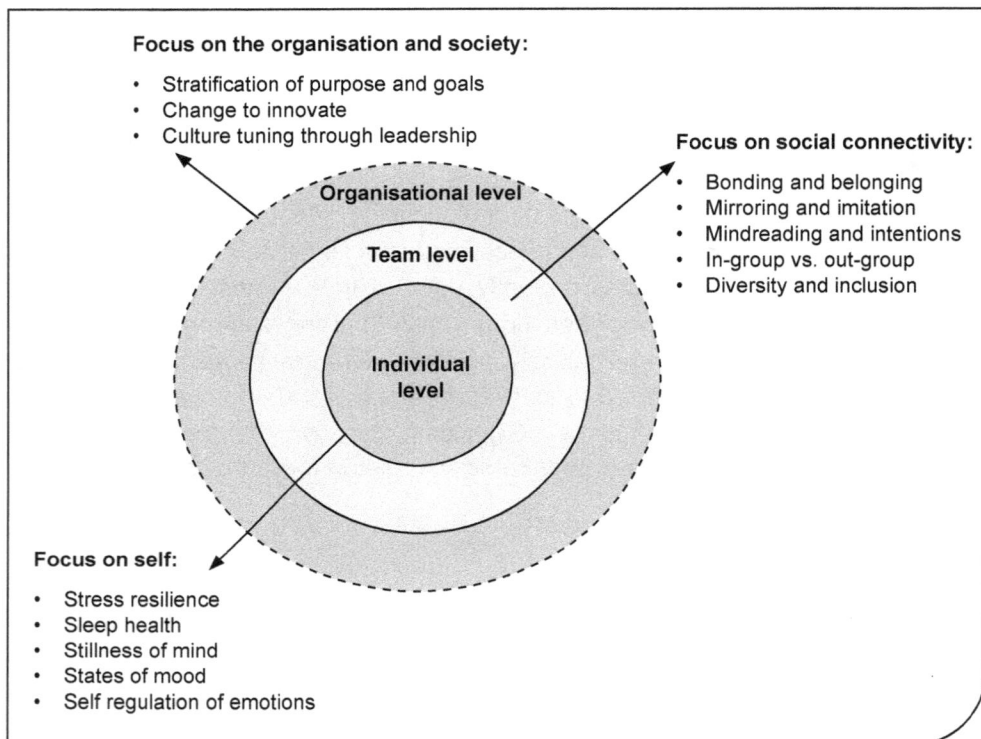

Focus on the organisation and society:

- Stratification of purpose and goals
- Change to innovate
- Culture tuning through leadership

Organisational level

Team level

Focus on social connectivity:

- Bonding and belonging
- Mirroring and imitation
- Mindreading and intentions
- In-group vs. out-group
- Diversity and inclusion

Individual level

Focus on self:

- Stress resilience
- Sleep health
- Stillness of mind
- States of mood
- Self regulation of emotions

Figure 12: The NeuroCapital Framework

The demands of our world at work can make us lead surface-level lives in which we are not able to take a moment to rest, let alone check in with ourselves or others. To leverage NeuroCapital at work, we need to actively deploy new behavioural strategies, learn new habits and unlearn some behavioural strategies and habits more effectively.

6.1 Stress: manage it and build resilience

"Every stress leaves an indelible scar, and the organism pays for its survival after a stressful situation by becoming a little older."

Hans Selye[78]

At a Glance

Stress is our response to a demand for change. Not all stress is bad. It is even essential for us to learn and grow at work. This section focuses on why stress was never meant to be experienced as chronic and elevated. Secondly, how we attend to perceived stress has radically different results in our brain and body. Thirdly, learning to stress right can help to eliminate the effects of stress and put our best-selves back in the driving seat. I conclude with some Neuro-hacks for quick stress relief at work.

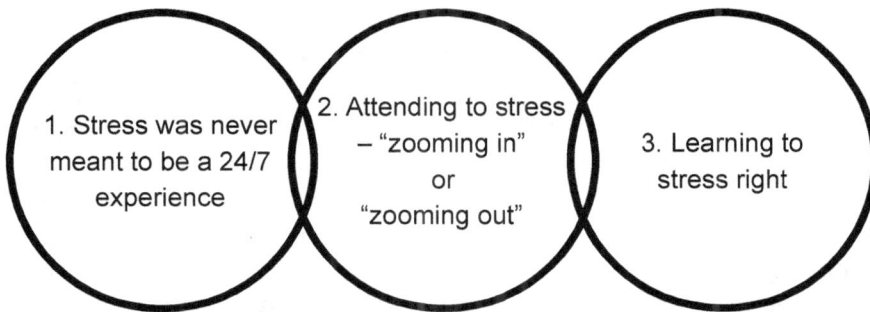

1. Stress was never meant to be a 24/7 experience

2. Attending to stress – "zooming in" or "zooming out"

3. Learning to stress right

1. Stress was never meant to be a 24/7 experience

Stress resilience in an exponentially changing world is a crucial concern for companies today – but how to stress less?

Believe it or not, stress is a normal set of psycho-physiological responses to a perceived threat. This is commonly known as the body's fight-or-flight response or stress response, which evolved when we as humans first prepared the body for physical activity. The stress response is the body's natural way of protecting you. The stress response can save your life – this is good stress (commonly called EU-stress), but it is a bad thing when we start feeling stress about non-life-threatening situations (so-called distress). We need to differentiate between 'good' stress and 'bad' stress; a certain level of stress keeps us going, that is EU-stress, without which we might never get off the couch.

In a state of high stress, the brain invokes the limbic system into hyperactivity. Hans Selye, the pioneer of stress research, had an exact description of the stress response. He described a sequence of neurophysiological changes that the body undergoes in response to injury, possible threat of harm, or life's minor ordeals. He proposed that the stress response is a universal reaction by the body to threats and dangers of all sorts, ranging from burns, blisters and bacteria threatening the body, to being rejected by our nearest and dearest or kicked off the project team at work. In short, when one perceives an event as a stressor, the brain sends a hint to the hypothalamus to secrete a substance called CRF, or "cortico-releasing factor". CRF travels through a particular gateway to the pituitary gland, where it triggers the release of ACTH (adrenocorticotrophic hormone), opioids and endorphins.[79] Seemingly, early in evolution this brain alarm went off when a mammoth or sabre-tooth tiger appeared. In modern times, a meeting with the auditors or the human resources department will do.

Driven by adrenaline and cortisol, the stress response prepares us for fight, flight or freeze responses, and at the same time, it dulls the pain response. From a functional-anatomical perspective, a hyperactive limbic system is an inflamed limbic system (filled with inflamed emotions), just like excessive physical training can cause a swollen knee that hurts at the slightest touch. In sum, whether physical or mental in origin, perceived pain registers in the brain via a system that can dampen its signals. The relief of pain is built into the brain's design, but this fight/flight/freeze response is supposed to be a swift reaction to get you quickly out of danger, not a 24/7 standby response. When animals escape they come right out of fight-or-flight mode and into "rest-and-digest" mode, where the parasympathetic nervous system is working to replenish their resources. On the contrary, it looks like us humans are stuck in emergency gear, being constantly vigilant against perceived threats.

Fear is the mother of all stresses. Research shows that even the plants and the water around us are polluted by our stress vibrations. For starters, chronic sustained stress makes us implicitly (i.e. not consciously) look more at angry faces. Moreover, during stress, that sensory short cut from the thalamus to the amygdala becomes more active with more excitable synapses, and we know the resulting trade-off between speed and accuracy. Compounding things further, brain chemicals (mainly cortisol) decrease activation of the (cognitive) medial PFC during processing of emotional faces, thereby reducing the accuracy of reading emotional expressions.[80] These are all symptoms of an over-stimulated central nervous system; in other words, we get trapped in anger = fight; anxiety = flight; depression = freeze.

The fight-or-flight response involves many real and demanding physical processes, and it takes time for the effects of these to wear off. The adrenalin rush alone takes a while to leave the bloodstream; we have tense muscles and knots in our stomach. But that is not all. There is also the increased emotional sensation. When you are primed to be terrified or angry, you cannot just switch off in an instant, and as we know, this frequently ends up being directed at less deserving targets. Tell an incredibly tense person to relax or calm down and see what happens.

Allostatic load is the expression used to describe the wear-and-tear effects of prolonged stress on the body. A high allostatic load leads to high blood pressure, impaired immune function, loss of brain function in the memory circuits of the brain, and growth of the fear circuitry in the brain. However, one's allostatic load can be contained, and a state of 'flow' can be induced through brain-based principles and behaviours.

Chronic elevated stress selectively impairs brain regions responsible for problem-solving, memory and emotional regulation (the amygdala, the hippocampus and the VL PFC). The problem occurs partly because of the interaction among three experiences: (i) chronic elevated stress negatively impacts on sleep and positive affect; (ii) poor sleep makes stress and positive affect worse; and (iii) negative affect tends to make sleep and stress worse. When we have high levels of all three experiences, the interaction can result in overly high stress experiences with substantial cognitive impairment, impacting on basic perception as well as judgement and decision-making.

This prolonged cognitive load leads to elevated stress and limbic system activation – the amygdala and hippocampus with spiked cortisol release – which sends us into a downward spiral of despair (see Figure 13). It reduces working memory, increases pessimism, reduces insight and reduces verbal fluency. Then consider the role of attention in all this; our attention is on surviving, not creating an outstanding client proposition. We become all or nothing thinkers and doers!

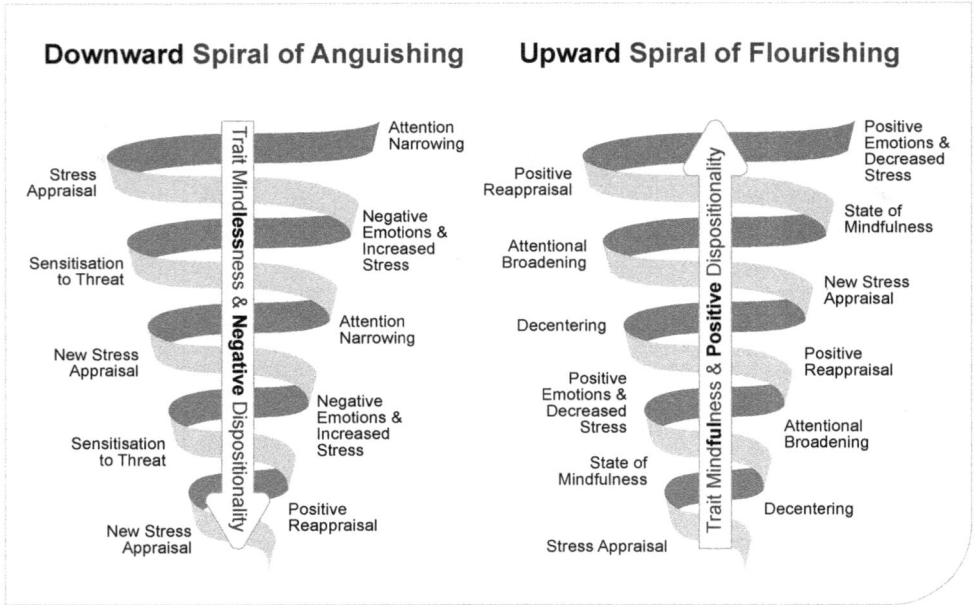

Figure 13: Downward and upward spirals (Source: Garland)[81]

2. Attending to perceived stress – zoom in or zoom out

Stress is not just a dilemma of too much going on. How we attend to our emotions and feelings is the critical element of our addiction to narrow focus. We've become habituated to stress, but maintaining a tense, emergency mode of attention tires us out and we need another cup of coffee to gather the energy to keep paying attention – or a cigarette or a glass of wine to relieve the tension of narrow focus.

Under stress we experience a 'threat state' where our attention narrows (see Table 4 below). Narrowed focus occurs when our well-being is threatened – a reflexive response to perceived fearful situations.

Table 4: Threat vs. reward states

Threat or Away State	Reward or Toward State
Reduced working memory and verbal fluency	Increased cognitive resources
Generalising threats	Fewer perceptual errors
Increased pessimism	Increased optimism
Narrow field of view	Wider field of view

Chronic narrowed attention eventually spikes stress levels. Even if we are the carefree type, stress can and does accumulate to levels that produce symptoms of disorder and disease (although we often do not recognise them as being caused by stress). Preventing the diffusion of stress and causing its accumulation, narrowed focus actually makes us less productive over the long haul. People who complain of an inability to concentrate, listlessness, low productivity and depression often find these problems resolved when they learn to shift out of narrow attention. Stress also generates physical symptoms, like an acidic stomach, sleeplessness or jaw tension.

On a psychological level, when we are stressed and remain in narrow focus, fear and anxiety play an exaggerated role in our minds and adversely colour our perceptions of the world around us. Although we may not realise it, narrow focus and the resultant stress that we bottle up inside keep us emotionally numb, blocking many feelings from our awareness. We miss out on rich experiences of smell and taste, pleasant physical sensations, and deep feelings of joy and sadness. Ironically, and tragically, although this constant narrow focus is how we attempt to connect emotionally with other people, and with experience itself, it is exactly the wrong way for making these kinds of connections.

3. Learning to stress right

Not all stress is bad. There is stress that we love. We love stress that is mild and transient and occurs when we are engaged, engrossed, challenged, being stimulated or playing. The core of psychological stress is loss of control and predictability when we are challenged by the unexpected – an off-the-cuff sarcastic comment or a client's cancellation of a critical project. The inverted U-curve of performance (or the optimal arousal curve) explains the relationship between cognitive demands, arousal or stress, and level of performance (see Figure 14 for the inverted U-curve). Optimal performance is achieved at the peak of the curve. This midpoint is characterised by a mentally stimulating state. Below the midpoint, performance declines as a result of insufficient arousal. Above the midpoint, arousal builds to levels that induce stress and anxiety caused by the task being perceived as beyond the capability of the individual. Some of us thrive amidst these challenges; we experience being at the top of the arousal curve or in a state of 'flow'. Others become frazzled or choke under the perceived pressure, which leads to performance decrements – also known as power stress.

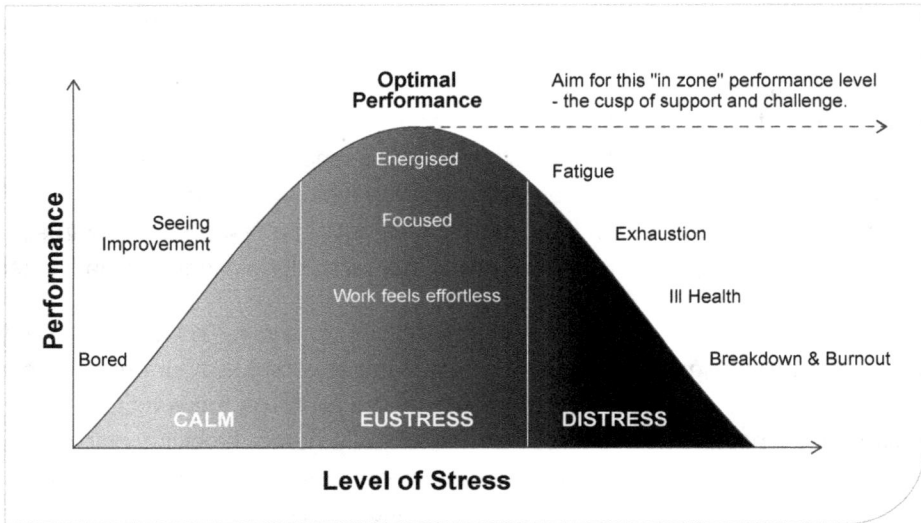

Figure 14: The inverted U-curve of performance (Source: Yerkes & Dodson)[82]

From a neuropsychological perspective, the mind-state that has been shown to be associated with consistently raised dopamine levels is one characterised by purposefulness, a sense of achievement or anticipated achievement, a sense of belonging and a sense of control. This is also called the Goldilocks Syndrome, where synaptic firing in the PFC depends on having just the right levels of catecholamines, dopamine and norepinephrine.[83] Not enough firing equates to a lack of engagement/ motivation. With too much arousal, stress is triggered or the brain's electrical breaker is flipped, resulting in a power outage.

#Goldilocks = Optimal motivation, concentration, working memory and other prefrontal cortex executive functions.

As mentioned before, excessively raised levels of noradrenaline and dopamine disrupt one's /executive function (the smart intuitive brain) The disruption of executive function results in a shift from reflective/insightful prefrontal activity to the reflexive fight-or-flight activity limbic system (impulsive brain) and finally automatic survival activity (the instinctive brain). Figure 15 provides gives a a graphic overview of how too much perceived stress makes us tip over into dysfunction.

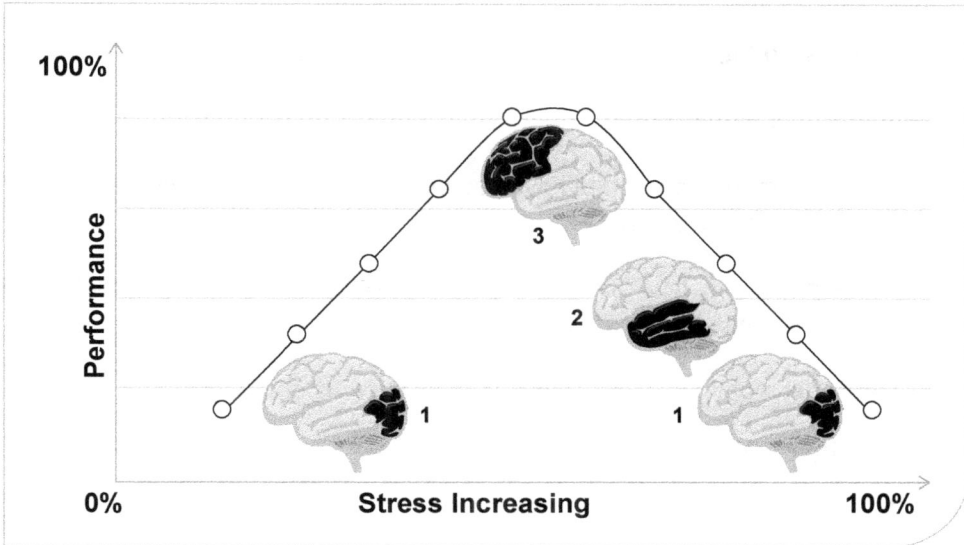

Figure 15: A neural lens: The stress vs. performance curve

(3) Cortical Brain = the executive functions (thinking, learning, language and inhibiting) or the intuitive clever brain

(2) Limbic Brain = the emotional system (attachment and emotional development) or the impulsive brain

(1) Brainstem = the instinctive brain (sensory motor input and survival) or the automatic brain

In a nutshell:

•   Chronic elevated stress (social and physical) damages many **core circuits**, over-activates one's limbic system and depletes one's **working memory**.

•   Stress enhances **negative emotional** memories.

•   Stress disrupts **neutral** memories.

•   Stress results in **memory loss** for **all** except the negative.

•   Stress leads to an **elevation in cortisol and heart rate** – it puts people in a sub-optimal brain state and reduces their capacity to change and learn (#AngerManagement@Work).

 # Neuro-Insight: Measure your perceived stress

Reflect on these questions:

- In the last month, how often have you been upset because of something that happened unexpectedly?

- In the last month, how often have you felt that you were unable to control the important things in your life?

- In the last month, how often have you felt nervous and "stressed"?

NOTES

_____

_____

_____

_____

_____

_____

_____

#Neuro-hacks to develop your stress resilience

- **Rehabilitate** your chemical system – it's all about balance. If you want to persuade your body's pharmacist – the hypothalamus in your mid-brain – into manufacturing all those delightful stress reductive natural chemicals, you must cut out the junk and maintain a well-balanced and varied diet. This will keep your blood-sugar levels steady and provide the essential building blocks for the natural internal uppers, downers, brain enhancers, and person connectors.

- Create sustained levels of oxygen **(more movement)** and glucose (low GI) – the brain always performs sub-optimally without this duo! Do short **bouts of exercise** during the workday as it decreases the impact of stress on the brain.

- **Down-regulate distress** by building perceived control into your life through exercise, nutrition and sleep hygiene. Exercise profoundly reduces the effects of chronic stress – brain performance enemy number one.

- Facilitate a shift in **cortical blood flow** to the frontal regions – through **breath control and mindfulness practice.**

- **Reframe:** When we unwisely exaggerate and minimise the daily pains and pleasures of life, we experience hormone-related anxiety, guilt-ridden stress and feel our lives are momentarily spinning out of control with chaos. Since each event in life has both positive and negative components and repercussions, anytime we misperceive such an imbalance, we become stressed. Stress happens when we are looking at only half of an emotional equation, not the ordered whole.

- Learn to shift into and **maintain optimal brain states**, where you gain the capacity to regulate your own physiology and enhance your performance and that of those around you. This helps avoid negative reactivity in difficult situations. Consider neurofeedback training to track your **biomarkers** to stress sensitivity and get to grips with your stress levels.

- Plan complex, strategic tasks first thing in the morning when the **inhibitory cortex** (the brain's braking system) **is fresh.** The brain region which governs inhibition is notoriously susceptible to fatigue, as controlling impulses is hard work.

- Check-out from your work-life and **get comfortable with being rigorously unproductive** once in a while. This also includes falling out of love with your phone – a love affair that often leaves us stressed-out, distracted and feeling inadequate.

Conclusion

Stress is the mother of all disease – physical, emotional, social, mental and spiritual. It really does not matter by whatever other names you call your problem – irritability, temper, frustration, tension, nervousness, hopelessness, sadness, lack of motivation or concentration – it still boils down to your perception of being trapped in instinctive survival and impulsive mindsets.

All work and no play will not build your stress resilience. Changing the default setting of your brain and nervous system from the instinctive, impulsive state to the intuitive poised state is the single most important strategy to cultivate your own NeuroCapital. This takes practice and kindness to self. Learning to deal with stress before it deals with you can prevent disease, optimise health and improve performance at work.

6.2 Sleep health

"The best bridge between despair and hope is a good night's sleep."

Matthew Walker[84]

At a Glance

Good sleep hygiene with a regular sleep pattern affects all your brain's information processing. It is a foundational driver of brain and body functioning. Sleep is NOT a choice. The focus in this chapter is on the impact of sleep deprivation, the types of sleep, what the brain does during sleep and a few neuro-hacks on how to join the sleep revolution.

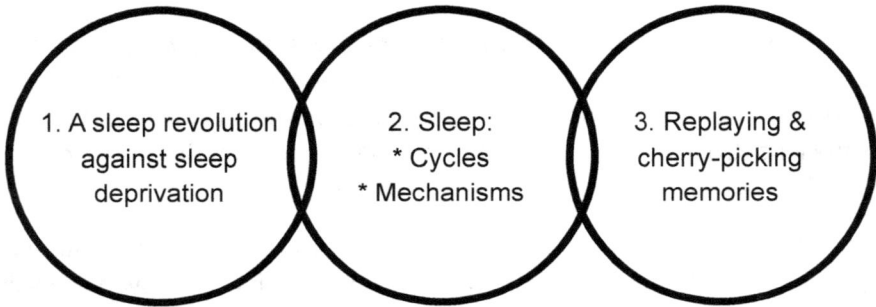

1. A sleep revolution against sleep deprivation	2. Sleep: * Cycles * Mechanisms	3. Replaying & cherry-picking memories

1. A sleep revolution against sleep deprivation

We may be what we eat, but we are also how we sleep. Increasingly, sleep deprivation is used as a badge of honour in a supercharged world of work, and "Why am I so tired?" one of the most used phrases that we think and talk about. The role of sleep in our lives is being rediscovered and researched like never before, yet the notion that there is more to sleep than just switching off is ancient. The quote: "Even a soul submerged in sleep is hard at work and helps make something of the world" has its origins in work by Heraclitus, a Greek philosopher who lived in 500 BC.[85] Fast-forward to the 21st century, where we suffer from the red-eye effect and often have a short fuse.

Sleep is indispensable for our nervous systems to work properly; too little sleep leaves us drowsy and incapable of concentrating the next day. Inadequate sleep selectively impairs those areas of the brain that are important for problem-solving, creativity and emotional regulation. Less dangerous cases of sleep deprivation typically involve short-temperedness, moodiness, illogical thinking and irrational behaviour. This happens because activity in our prefrontal cortex – the 'CEO of the brain' that rules rationality and logical thinking – is dampened. If you often feel emotionally off-kilter or struggle with a short fuse, chances are you might manage your emotions much better if you were to get more sleep on a nightly basis.[86]

If sleep deprivation carries on, hallucinations and mood swings may develop. Sleep provides the neurons we've used (while we were awake) a chance to shut down and repair themselves. Without sleep, neurons may become so worn-out, depleted in energy or so polluted with by-products (neural debris) of normal cellular activities that they begin to malfunction. There is indeed power in slumber!

When we are sleep-deprived, the following happens to us:

- Our **cortisol levels increase** (the amygdala and hippocampus have the highest level of cortisol receptors in the brain), amplifying our perception of the negative in the environment. We become supercharged on the negative.

- We have difficulty processing **positive cues and signals around us** because the threat network is on red alert as it tries to help the tired, unalert brain fend off any danger.

- Insufficient sleep also means that we are not giving our brains time to integrate information in a meaningful way, and this sub-optimal functioning impacts our **creativity**.

- **Insulin regulation** is interrupted, which can lead to blood sugar issues and weight gain.

- Impaired sleep may **limit cognitive resources** to manage stress and increase hyperarousal. Memory becomes faulty.

- Sleep deprivation is the best interrogation strategy as the **brain's braking system** (VLMPFC) cannot regulate or brake when it is sleep-deprived. Thus, emotionally, we fly off the handle as the ability to control our impulses is reduced. Regulating emotions is a crucial strength for any person at any organisational level.

Jet lag or fatigued at work?

When we travel and pass from one time zone to another, we suffer from disrupted circadian rhythms, an agonising feeling known as jet lag. For instance, if you travel from Johannesburg to Kathmandu, you "lose" four hours according to your body's clock. You will feel exhausted when the alarm rings at 8 a.m. the next morning because, according to your body clock, it is still 4 a.m. It generally takes several days for your body's cycle to adjust to the new time.

Symptoms much like those of jet lag are common when you work nights or shifts. Because these types of work schedules are at odds with powerful sleep-regulating cues like sunlight, you will often become uncontrollably drowsy during work, and you may suffer insomnia or other problems when trying to sleep. Shift workers have a bigger risk of heart problems, digestive disturbances, and emotional and mental problems, all of which may be linked to their sleeping problems. The frequency and severity of workplace accidents also tend to increase during the night shift. Major industrial accidents are attributed partly to errors made by fatigued night-shift workers; one study also showed that medical interns working on the night shift are twice as likely as others to misinterpret hospital test records, which could put their patients at risk.[87] Shift-related fatigue might be reduced by using bright lights in the workplace, minimising shift changes, and taking scheduled naps.

2.   Sleep types, cycles and sleep mechanisms

Sleep types and sleep cycles

There are two main types of sleep, rapid eye movement (REM) sleep and non-REM sleep. REM sleep happens in roughly 90-minute cycles and alternates with four additional stages (stages 1-4, in order of increasing depth) identified as non-REM sleep. Slow wave sleep (SWS), the deepest of the non-REM phases, is described as high-amplitude, low-frequency brain oscillations, and is characterised by delta waves (measured by EEGs). Dreaming and sleepwalking can occur during SWS, and SWS is thought to be important for memory consolidation.

REM sleep, on the other hand, is a lighter state of sleep characterised by eye movements, decreased muscle tone (which inhibits the acting out of dreams), and low-amplitude, fast brain oscillations (theta and beta), resembling wakefulness. REM sleep is a neurophysiological state that is more comparable to wakefulness than non-REM states.[88]

More than 80% of SWS is concentrated in the first half of the night, whereas the second half of the night comprises roughly twice as much REM sleep than the first half. Healthy sleep consists of numerous stages, and we move through these stages four to five times during the nightly sleep cycle. In stage 5, we enter REM sleep, where dreaming takes place.[89,90]

Sleep mechanisms

Circadian rhythms are steady changes in mental and physical characteristics that occur over the course of a day (circadian is Latin for "around a day"). Most circadian rhythms are organised by the body's biological "clock". This clock is found in a part of the brain called the hypothalamus. Light that extends to photoreceptors in the retina (a tissue at the back of the eye) creates signals that travel along the optic nerve to the hypothalamus and eventually the pineal gland, which responds to light-prompted signals by switching off production of the melatonin hormone. Within the hypothalamus is the suprachiasmatic nucleus (SCN) – clusters of thousands of cells that receive information about light exposure directly from the eyes and control your behavioural rhythm. The body's level of melatonin typically increases after darkness falls, making us feel drowsy. The HPA axis (Hypothalamus, Pituitary gland, Adrenal glands) also governs functions that are synchronised with the sleep/wake cycle, including body temperature, hormone secretion, urine production and changes in blood pressure.[91] The Pineal gland, located within the brain's two hemispheres, receives signals from the SCN and increases the production of melatonin, which helps put you to sleep once the lights go off.

3. The brain replays to cherry-pick memories

As we sleep, facts, events and skills learned during the day continue to be processed. Memories become enhanced, stabilised and integrated with older memories, a process known as consolidation.

The brain also cherry-picks what we remember during sleep, resulting in sharper and clearer thinking. One of the processes is dreaming, which serves a number of important psychological functions (aiding memory formation, creative problem-solving), helping you find meaning in life events and imagining a different future. And, yes, everyone dreams. We spend about two hours each night dreaming but may not remember most of our dreams. Its exact purpose is not known, but dreaming may help us process emotions. Events from the day often invade our thoughts during sleep, and when we suffer from stress or anxiety, we are more likely to have frightening dreams. Dreams can be experienced in all stages of sleep but are usually most vivid in REM sleep. Some people dream in colour, while others only recall dreams in black and white.

Jessica Payne, a sleep expert at the University of Notre Dame and Harvard University, says that when we dream, the memory details that seem to get remembered best are often the most emotional ones. Payne and her colleagues found that when individuals are shown a scene with an emotion-laden item in the foreground – such as a wrecked vehicle – they are more likely to remember it than, say, palm trees in the background, especially if they are tested after a night of slumber.[92] Rather than maintaining scenes in their entirety, the brain apparently restructures scenes to remember only their most emotional and most essential components, while allowing less emotional details to deteriorate.

Measurements of brain activity support this concept, revealing that brain regions linked with emotion and memory consolidation are sometimes more active during sleep than when awake.[93] It makes evolutionary sense to selectively remember emotional information – our ancestors would not want to forget that a snake was in a specific location or that someone in the tribe was particularly mean and should be avoided – so it turns out that memories are not only about remembering the past, but are also about being able to anticipate and predict multiple possible futures.

How does this happen? The hippocampus replays experiences during quiet rest periods, and this replay benefits subsequent memory. This means that the re-expression of hippocampal activation during sleep reflects the offline processing of memory residue, which in turn leads to the strengthening of network connections in the brain, resulting in improved memory performance. Thus, after the day's events during subsequent sleep, there is a reactivation or 'replay' of this hippocampal activation, as if the brain is reprocessing recently learned information.[94]

Long, uninterrupted sleep makes learning count because learning is a vastly unconscious process and really occurs after the learning session during sleep. This consolidation of memories fixes them in the brain so we can retrieve them later. Sleep-enabled memories may help us produce insights, draw inferences, and foster abstract thought during waking hours[95], thus moving us from DUH to AHA moments!

From a practical standpoint, sleep research continues to show the importance of sleep to the brain – especially if you are learning new information and if you want that information stored in long-term, complex networks of neuron branches.

? # Neuro-Insight: How sleepy are you? Measuring your propensity to nod off

The daytime sleepiness scale provides a measure of a person's general level of daytime sleepiness, or their average sleep propensity in daily life. It has become the world standard method for gauging daytime sleepiness and was developed by Dr Murray Johns (called the Epworth Sleepiness Scale). He created the scale at the Epworth Hospital in Australia as a baseline for determining how sleepy a person is.[96]

Use the following scale to choose the most appropriate number for each situation. Then, fill in your answers in the table below:

Table 5: The Epworth Sleep Scale (see https://epworthsleepinessscale.com/about-the-ess/)

0 = would *never* doze or sleep. 1 = *slight* change of dozing or sleeping 2 = *moderate* change of dozing or sleeping 3 = *high* chance of dozing or sleepng		
	Situation	**Chance of dozing or sleeping**
1.	Sitting and reading	
2.	Watching TV	
3.	Sitting inactive in a public place	
4.	Being a passenger in a motor vehicle for an hour or more	
5.	Lying down in the afternoon	
6.	Sitting and talking to someone	
7.	Sitting quietly after lunch (*no alcohol*)	
8.	Stopped for a few minutes in traffic while driving	
	TOTAL:	

Scoring key:

The Epworth Sleepiness Scale is used to determine the level of daytime sleepiness.

A score of 10 or more is considered sleepy – and not fit to drive a car!

A score of 18 or more is very sleepy.

If you score 10 or more on this test, you should consider whether you are obtaining adequate sleep, or need to improve your sleep hygiene.

NOTES

_____

_____

_____

_____

_____

_____

_____

_____

_____

_____

_____

_____

_____

_____

_____

_____

_____

Neuro-hacks to leverage the power of sleep

- Understand the **interrelatedness** of sleep, stress and mood, and manage your sleep/wake cycles.

- Adopt "**rhythmic living**" as your health and wellness payoff line, for example, decide on a set time to go to bed every night which reduces mental volatility – a lower glucose metabolic rate, to be precise.

- Avoid working on **light-emitting devices** when in bed – light nights on PCs or a TV are not good. There is evidence that even low levels of light from a computer can change your circadian rhythm.

- Avoid having **emotional discussions** before going to bed – it activates cortisol release which inhibits sleep. Rather, laugh yourself to sleep, as humour enhances recall memory.

- Go to bed in a **state of grace** – it ensures good REM sleep. Adequate REM sleep has been correlated to **resilience**. Ensure at least eight hours of sleep during emotionally stressful times.

- Take **a power nap** – as little as a **seven minute nap** allows for the regeneration of creative brain circuits and improves one's mood.

- Prime your brain to **encode your life goals** by reviewing them before you go to bed. At night, the hypnagogic state (the point just before falling asleep) has "the Tetris effect", which means that the last thing you look at sticks or makes a lasting impression on your brain and begins to pattern your thoughts. Note that for priming to be effective, it needs to be congruent with goal-relevant intentions (like chief life goals).

- Create a work culture that is in **favour of sleep**. New recruits to an organisation easily fall into the cultural mode of behaviours of the existing work force. Unfortunately too many organisations still have cultures that encourage more of a workaholic and "I can do the red eye" approach that is linked to getting just a few hours of sleep.

Conclusion

Without sleep, you cannot form or maintain the brain pathways that let you learn and create new memories, and it is harder to concentrate and respond quickly. During sleep, we not only rest and restore strength for the next day, but sleep is also a vastly dynamic state that is essential for cognitive processes like memory consolidation, semantic integration, learning, and the processing of emotions.

Sleep offers us a recharge or renewal from our daily 'to do' lists, and allows us the timelessness and the opportunity to make connections that have eluded our conscious thoughts. We do not just sleep to rest, i.e. sleep does not merely provide rest that enables us to begin the next day more wakeful, attentive and alert. Rather, sleep and wake cycles are both highly active brain states.

Sleep also helps us consolidate memories, fixing them in the brain so we can retrieve them later. Sleep-enabled memories may help us produce insights, draw inferences, and foster abstract thought during waking hours. Sleep is NOT a choice.

6.3 Stillness of mind: switch to mindful awareness

"Education of attention is education par excellence."

William James[97]

At a Glance

There is a business case for stillness of mind which shows that the practice of silencing your mind and developing mindful awareness contributes to both stress reduction as well as enhanced social and cognitive engagement at work. Silencing the mind helps us to dissolve unprocessed emotional baggage that is holding us back at work and in life. This section focuses on: (1) the neuroscience research that validates why we should cultivate silencing our mind and cultivating mindfulness; (2) the two modes of experiencing the world, i.e. the story modes direct experience; and (3) working through the body to silence the mind through breath awareness and breath regulation.

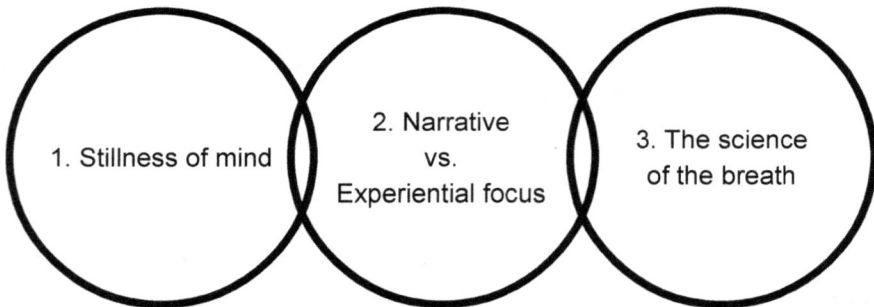

```
    ( 1. Stillness of mind )  ( 2. Narrative vs. Experiential focus )  ( 3. The science of the breath )
```

---

1. Stillness of mind to develop Mindful Awareness

You might roll your eyes at this one, but stillness of mind is a cornerstone of building brain health and your NeuroCapital. Stillness of mind, or mindfulness, also loosely known as self-awareness, means developing a complete and conscious awareness of yourself, others, your environment, and work.

The concept of mindfulness is often described in terms of its attentional and affective parameters. Mindfulness encourages a non-elaborative, non-judgemental, present-centred awareness in which thoughts, feelings or bodily sensations that arise in the attentional field are acknowledged and accepted as they are.[98] In other words, seeing reality as it is vs. how we want it to be.

A small but growing body of work in the field of Organisational Behaviour suggests that mindfulness is linked to better workplace functioning. Mindfulness appears to positively impact human functioning overall – it improves attention, cognition, emotions, behaviour and physiology. Although mindfulness is an individual quality, initial evidence suggests that it affects interpersonal behaviour and workgroup relationships. Mindfulness may improve relationships through greater empathy and compassion, which suggests that mindfulness training could enhance workplace processes that rely on effective leadership and teamwork.[99]

2. The neural footprint of mindfulness (Narrative Focus vs. Experiential Focus)

Mindful awareness works on similar brain centres as those affected by anti-depressants (serotonin increases with mindfulness practice). Mindfulness also works the brain the way a good workout regimen works the body – minus the buckets of sweat. Mindfulness enhances working memory capacity, which is vital for any corporate citizen.

Mindfulness practice is one way to integrate the various regions of the brain that facilitate psychological well-being. Not only does mindfulness practice have stress reductive effects, but it also enhances cognitive functions and resilience, which are vital at any organisational level. With mindfulness practice there is a beefing up of the middle prefrontal cortex (MPFC), which plays a vital role in integrating higher 'intellectual' brain areas with those in the more vulnerable 'emotional' areas. The brain builds synapses, synaptic networks and layers of capillaries, becoming thicker in two significant regions. One is in the prefrontal cortex, which is involved in the executive control of attention, i.e. deliberately paying attention to something. This change makes sense because that is what you are doing when you meditate. The second brain area that gets bigger is the insula, which enables you to become more self-aware and empathic.

Neuroscience-based research shows that people who are high in mindfulness have an improved ability to regulate their emotions (right and left ventral PFC activation). An additional benefit of mindfulness is that it enables easier adaptability to demanding situations and enhances emotional regulation – a skill that is especially important in the workplace. Mindfulness stimulates the activation of brain regions associated with self-monitoring and cognitive control. It decreases grey matter density in the amygdala, which plays an important role in anxiety and stress. We should therefore learn to be more in charge of the focus of our attention and emotionally regulate ourselves – feeling the right emotion at the right time in order to overcome perceived disturbances (such as corporate change or a horrible boss) – because the brain's braking system can then be effectively recruited when needed. A brain without brakes is not nice to be around!

Mindfulness promotes stronger activation in the temporal parietal junctures, another part of the brain tied to empathy.

The neural correlates of the process of mindfulness were recorded in a ground-breaking neuro-imaging study, where functional magnetic resonance imaging (fMRI) was used to examine the two neural modes of self-referencing: 'narrative' focus, or NF, and momentary experience, known as 'experiential' focus, or EF, as well as the neural systems supporting these 'modes' of awareness.[100] The study revealed that mindfulness meditation enables a decoupling between the narrative mode and the experiential mode. These two forms of self-awareness are habitually integrated but can be dissociated through attentional training. The decoupling of these two forms of awareness enables one to choose which mode is required for the task at hand.

The narrative focus equates to the conceptual world that is full of descriptions of events, memories, attitudes, and evaluations of people and things. The experiential focus is about tracking how one's body, thoughts and feelings change in an instant to that of a judgment-free awareness of current experiences and intentions. These two modes are completely different, have different neuroscientific underpinnings, and are anti-correlated. Thus, if the one is 'off', the other one is 'on'.

Mindfulness is about being able to be aware of which mode you are in and then being able to switch to the circuit that is most beneficial for the task at hand. These two modes are set out in Table 6 below.

Table 6: Two modes of self-referencing

Narrative Mode Circuit	Direct Experience Mode Circuit
Making a story about something Focus is through time (*past, present, future*)	Processing incoming data as they happen (*in real time*)
• **Circuitry involved**: Conceptual, memory (hippocampus) and limbic (*friend or foe decisions*) • Metabolically expensive in terms of brain resources	• **Circuitry involved**: Somato-sensory cortex and insula (*internal visceral experience*) • Metabolically stable on brain resources
Leads to fixation on story line, can lead to state of hyper arousal and even cognitive shut down	Leads to better self-regulation and the ability to take in more new data, which increases reflective thinking, enables new insights and reduces negative biological effects like hypertension

To 'fine tune' your direct experience mode:

- breathe in slowly through your nose and exhale all your breath;

- focus your attention on your body or one part of your body, such as your feet; and

- hold your attention on the direct experience and take in the data – do not process the data.

Examples of how to induce the direct experience mode:

- When you walk up the stairs, be aware of your breathing, your movement up the stairs, the feeling of your feet on the steps, i.e. be aware of the sensations.

- When you get to work, sit in your chair, close your eyes, focus on your breathing and pay your attention to your feet, i.e. be aware of the sensations.

- Other '1-5 minutes a day' examples: washing dishes, going for a walk, eating, working in the garden, waiting for a meeting to start, going for a smoke break, any exercise, a mindful shower.

3. The science of the breath

Conscious breathing builds the connections between the limbic system and the neocortex.

- Our breath is mainly automatic and unconscious; it is regulated by our instinctive and primitive involuntary or autonomic nervous system.

- By practicing conscious breathing, we gradually strengthen the connections between our unconscious, autonomic reactions and our voluntary or consciously chosen responses.

- This strong bridge between the primitive "lower" mind and the "higher" mind enables us to mend the split between raw reactions and wiser responses.

- Observing the breath allows us to take a deep look at the nature of mental formations such as fear, anger and anxiety.

Recent research by De Couck et al. entitled, *How breathing can help you make better decisions: Two studies on the effects of breathing patterns on heart rate variability and decision-making in business cases*[101], reported that:

- just two minutes of deep breathing with longer exhalation engages the vagus nerve, increases HRV, and improves decision-making;

- performing deep breathing exercises diminishes perceived stress after a challenging decision-making task; and

- performing deep breathing exercises improves decision-making task results (nearly 50% more correct problems).

(?) # Neuro-Insight: Mindful attention awareness

Reflect on the behavioural pointers below: do they describe your state of attention awareness of, ongoing events or not?

1. I do jobs or tasks automatically, without being aware of what I'm doing (yes/no/maybe).

2. I find myself listening to someone with one ear, doing something else at the same time (yes/no/maybe).

3. I drive to places on "automatic pilot" and then wonder how I got there (yes/no/maybe).

NOTES

_____

_____

_____

_____

_____

_____

_____

"It is impossible for a man to be cheated by anyone but himself."

Ralph Waldo Emerson[102]

#Neuro-hacks for cultivating stillness of mind

The old Zen saying, "You cannot wash off blood with blood", refers to the notion that it is difficult to control thoughts with other thoughts. This saying implies that the way to control the mind is through the body. The primary way to achieve this control of the mind is through breathing and posture.

- **Learn to shift from narrative mode to direct mode.** This will enable you to reflect in a deeper manner as you will have more access to mental bandwidth when in a calmer state.

- Make time and space in your day for **sensitive reflection, knowing when to switch from analytical (hyperfrontality) to quiet mode (hypofrontality)** by taking a walk, listening to music, silencing the mind, or journalising. **This hypofrontality enables a broader 'field of view' instead of a myopic focus.**

- Develop mindful attention awareness. The best avenue to get to observer status is to **slow down our thought processes**. This is done through breath awareness and breath regulation, not through talking or thinking. Your breath cycles are a function of a two-gear system – sympathetic vs. parasympathetic. Breathing awareness and breathing regulation assist with dissolving unprocessed emotional memories, also called *#Brainflossing*.

- **Make better breathing your middle name.** For a start, taking longer exhalations is an easy way to hack the vagus nerve, which counteracts fight-flight-freeze stress. An inhalation-to-exhalation breathing cycle per minute of a ratio 4:8 (four-second inhalations and eight-second exhalations) achieves this counteraction.

- **Self-awareness** entails paying attention to inner physiological cues. However, if you are in cognitive overload and trapped in autopilot tendencies, you will not be able to sense these cues or gut feelings. These are also called **somatic markers**, the purpose of which is to simplify decision-making by directing attention to better options.

- **Learn to surf the perfect storm**. Neuroscientist Amishi Jha's formal study on mindful attention in the US Marines showed that 12 minutes of mindfulness meditation per day increased resilience compared to Marines who did less than 12 minutes or no meditation.[103]

Neuro-Insight: A mindful breathing practice

It can be tricky to adopt stillness of mind and mindful attention awareness as new habits, but there are simple and smart ways to begin to make these practices routine. One of the most important things to sustain in the workplace today is how to focus. Mindful breathing can help you strengthen your focused attention.

If you do this exercise, say 10 minutes before you go to work or at your desk, you are changing your brain and heightening your ability to concentrate hours later. Mindful breathing practice:

- Sit upright, close your eyes and bring your attention to your breath.

- Do not try to control your breath – just let it be natural and easy, but be aware of it. Notice the full inhalation and the full exhalation.

- See if you can feel it coming and going through your nostrils or feel the rise and fall of your belly.

- When you notice that you have been distracted, simply start with the next breath.

- Tune in to any sensation any way you can. Take in sensory data. Be fully aware of the breath. Just keep your attention anchored there. Do not aim to process the data.

- Keep breathing in and breathing out. Whenever your mind wanders, just bring it back to your breath. Watch the full inhalation and the full exhalation. Stay with the breath. Use it as your anchor for attention.

- Try it on your own for a few minutes.

NOTES

_____

_____

_____

_____

_____

_____

_____

It's really so simple and in some ways so hard, because the mind wants to wander. In a way, the basic movement silencing your mind is about anchoring your attention, keeping it there, noticing when your mind wanders because it's going to, bringing it back and starting over. What we find is that if you can keep doing this, and the longer you stay with your breath, the more relaxed your body becomes. It's a side effect of that full attention and letting go of all those worries that keep us on edge and distracted.

Conclusion

Stillness of mind practices equate to good brain hygiene and act as a type of "brain flossing" that is as important as brushing our teeth. Stillness of mind sounds so simple but, in some ways, it's hard because the mind wants to wander. Learning to take purposeful pauses with breath regulation is a basic repetition to develop mindful attention awareness. A state that can become a trait!

6.4 States of mood

"I think of a pessimist as someone who is waiting for it to rain.
And I feel soaked to the skin."

Leonard Cohen[104]

At a Glance

Our mood does not just happen by accident. Instead, it is first created by a thought, sometimes a conscious one, while at other times it may be a repetitive and less conscious core-belief system. The focus in this section is: (1) the purpose of mood; (2) how to neutralise our mood; and (3) how the positive neuroscience of gratitude practice can be applied to cultivate an even mood state.

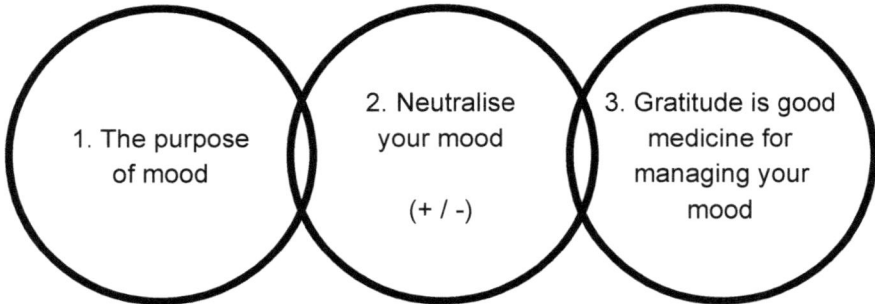

1. The purpose of mood

2. Neutralise your mood

(+ / -)

3. Gratitude is good medicine for managing your mood

1. The purpose of mood

Mood is an internal, subjective state, but it can often be inferred from posture and other behaviours. We can be sent into a mood by an unexpected event, from the happiness of seeing an old friend to the anger of discovering betrayal by a partner. We may also just fall into a mood.

In psychology, a mood is an emotional state that is extended over time, for example, anxiety is a mood and fear is an emotion. In contrast to emotions, feelings or affects, moods are less specific, less intense and less likely to be provoked or initiated by a particular stimulus or event. Moods are typically described as having either a positive or negative valence. (Valence is Latin for 'strength'.)

In other words, people usually talk about being in a good mood or a bad mood. Mood also differs from temperament or personality traits, which are even longer-lasting. Nevertheless, personality traits such as optimism and neuroticism predispose certain types of moods. Long-term disturbances of mood, such as clinical depression and bipolar disorders, are considered mood disorders.

Neuroimaging studies show that when it comes to mood:

- positivity widens our attention and our receptiveness to the new and unexpected[105, 106], while pessimism narrows our focus enabling survival responses;

- when people feel positive they are more creative and broaden their visual attention[107, 108]; and

- a positive mood facilitates memory for neutral and positive information and increases verbal fluency, creativity and problem-solving.[109]

A poorly-managed negative mood is not good for your health. Negative attitudes and feelings of helplessness and hopelessness can create chronic stress, which upsets the body's hormone balance, depletes the brain chemicals required for happiness, and damages the immune system.

 # Neuro-Insight: How many of these statements apply to you?

1. In uncertain times I usually expect the best **(yes/no/maybe)**

2. It's easy for me to relax **(yes/no/maybe)**

3. If something can go wrong for me, it will **(yes/no/maybe)**

NOTES

_____

_____

_____

_____

_____

_____

However, in nature everything serves a purpose, otherwise it would not have survived. The same goes for mood; negative and positive moods both serve a purpose. The arousal of the autonomic nervous system (ANS) via our neuroception largely determines whether we will have an imbalanced mood or not. When the parasympathetic nervous system (PNS) is activated, it results in feelings of optimism and safety. The activation of the sympathetic nervous system (SNS) results in attention narrowing, with a decrease in the quality of thinking.

The brain closes the nonessential neural pathways and thinking becomes reactive and inflexible. The Positive and Negative Emotional Attractors (PEA and NEA) are two attractors that represent two key drivers. These are the basic need to survive and the need to thrive.[110] The PEA will elicit the arousal of the PNS, while the NEA will arouse the SNS. Thus, PEA and PNS act as a positive force in influencing our thoughts and subsequent behaviour, like positive mood, vision, optimism and hope. The message is to intentionally tune into the positive. As Boyatzis[111] argued, when people are in a positive emotional state, they are "more perceptually open and accurate in perceptions of others". On the contrary, negative emotional states (NEA and SNS) are said to be linked to human survival, particularly to defend against threats.

Prior research on the intensity of moods suggests that in order to move a person from a negative emotional state to a positive emotional state, the intensity of the emotion must be reduced to reach a tipping point.[112] Although contested, the essence of mood ratios are useful, for example, to maintain a positive mood, keep a minimum 3:1 positive to negative ratio, and a 6:1 ratio to flourish.[113]

Mood is not just an individual state of mind. For example, Sigal Barsade's discovery of the "Ripple Effect" shows how moods can transfer among groups and influence their behaviour, emotions, and even personal opinions.[114] A positive and upward spiral team mood can be developed by setting up 'pro-cues', i.e. by helping individuals set up 'cues' in their environment to induce a more positive toward state in their daily work.

2. Neutralise your mood (see the good in the bad and the bad in the good)

Essentially, we need to learn how to see the good in the bad and the bad in the good. This is emotional literacy which forms the foundation of emotional stability, insight and a meaningful life. When an emotional balance sheet is drawn up, the bottom line should be about zero. Dr John Demartini's[115] method of neutralising emotional 'charge' is a break-through (and a beautiful) way of learning to do this.

To be in this neutral state where we are content and calm requires action and dealing with lopsided perceptions that drive volatile moods. Some of these perception errors that lead to feeling blue, negative and having volatile moods include:

- not seeing that the good and the bad in the world reflect the balance of order (#Chaos Theory);

- not seeing that energy is never lost – it only changes form;

- ignoring the fact that light casts a shadow;

- ignoring the fact that darkness suggests light;

- exaggerating the good in the good;

- exaggerating the bad in the bad; and

- not seeing that the good and the bad you see in the world and in others are reflections of yourself.

When good and bad are balanced out, the result is that you do not desire much, nor do you resist much. However, this does not mean that you are brain-dead. The neutral mood state is not empty – it's filled with poise and inspiration. The closer you are to the point of zero emotion, the closer you are to ultimate goodness, or pure potentiality. Paradoxically, that puts you in the position of maximum power to create and to change for the better.

"I regret nothing (Je ne regrette rien)"

Edith Piaf[116]

The French singer Edith Piaf might well have regretted nothing, but most of us experience a lot of regrets. Regret is a complex emotion that encapsulates the feeling of disappointment for a decision that was suboptimal (with the clarity of hindsight). It might be feeling sadness or anxiety at having failed at something or having made a poor decision. Regret negatively taints our mood, but there is a way to dissolve regret or any other emotional charge that affects your mood.

The exercise below is a sure way to neutralise your mood.

Neuro-Insight: See the good (+) in the bad (-)

Write down something that you regret. Then reflect and write down how this was a blessing to you or how this incident has served you.

When you replace "Why is this happening to me?" with "What is this trying to teach me?" Everything shifts and miracles happens at work!

NOTES

_____

_____

_____

_____

_____

_____

_____

_____

"You will continue to suffer if you have an emotional reaction to everything that is said to you. True power is sitting back and observing things with logic. True power is restraint. If words control you, that means everyone else can control you. Breathe and allow things to pass."

Warren Buffet[117]

3. Gratitude is good medicine for managing your mood

"Gratitude can transform common days into thanksgivings, turn routine jobs into joy, and change ordinary opportunities into blessings."

William Arthur[118]

If you have been running in supercharged mode for a while, you will inevitably fall into some kind of negative mood state, due to the inevitable depletion of the body and brain chemicals you need for happiness, energy and connection. Having an attitude of gratitude in this state sounds bizarre, but it is medicine that works and you do not need a script.

Saying thank you and receiving a thank you feels good. In a state of grace, it is difficult to be your worst self. When you are grateful for things when they are not as you want them to be, you see the perfection right here, right now. Neuroscience research findings contribute to this blissful subjective experience. As shown by Kini[119], expressing gratitude can have lasting effects on a brain region that is important in emotional regulation and social reward. Gratitude practice increases activity in the MPFC, which is associated with self and other awareness. Thus, you have more tolerance for self and others.

Gratitude also enhances physiological bliss. Figure 16 illustrates the heart rate variability pattern of frustration (top), which is characterised by its random, jerky pattern. Sincere, positive feeling states like appreciation (bottom) can result in highly ordered and coherent Heart Rate Variability (HRV) patterns, which are generally associated with enhanced cardiovascular function and mental/emotional balance.

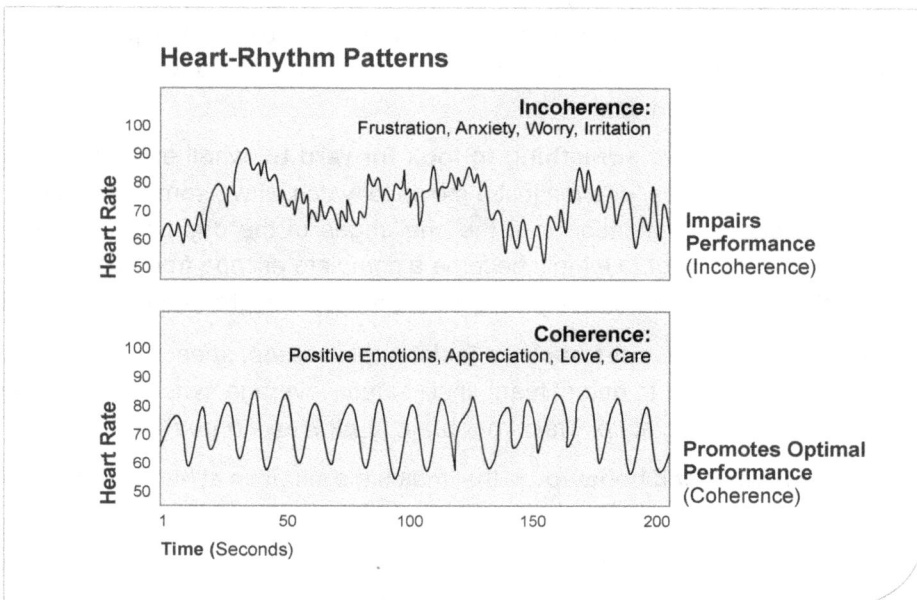

Figure 16: Heart rate variability – a key measure of mental and emotional balance (Source: The Healed Tribe)[120]

"I didn't think of all misery but of the beauty that still remains."

Anne Frank[121]

#Neuro-hacks to bring mood states into balance

- Become cognisant of the **innate negativity bias** in the brain and **deliberately** focus on deploying **positivity in all interpersonal interactions.**

- When it comes to mood, small things matter: **a smile** in the first five minutes sets up the brain for a cascade effect of good mood.

- Do **positive affirmations**, ideally in written form, as they reduce negative outlooks.

- Check in with people who give a **balanced viewpoint** – see what positives they see in the situation. What is their perception of the situation?

- Determine what could **go right** with a decision as well as what could **go wrong**. This also reduces emotional volatility.

- Use power posing when you are feeling very anxious. **Power posing** can increase the subjective perception of power and possibly increase testosterone.

- Be aware of your own mood and change it if it is not suitable. One way to do that is to change your **facial expression**. The facial feedback hypothesis states that our facial expressions impact our emotions. **Intentionally smiling** leads to feeling positive emotions.

- **Expectations; have something to look forward to,** whether it's a holiday, a catch-up with friends, or even just a movie to watch. Have something you enjoy that you can shift attention to if the challenges of the day become a bit too much. Be careful not to let this become a complete escape from reality though. Sometimes if everyday life is getting too hard, then it might be time for a change.

- **Tune into affective cues such as facial expressions**, speech pace and tone, and body language to detect fears, instil safety, leverage synchrony and evoke a higher performing state. Train your brain to **pick up on positive cues.**

- **Display moods and behaviours** that match the situation at hand, with a healthy dose of optimism mixed in.

- **Respect how other people are feeling** – even if it is glum or defeated. Model what it looks like to move forward with hope and humour.

- **Food and mood**. There is a connection between the two – adults who consume more omega-3 fatty acids, calcium, folate and vitamins B1, B2, B5, B6, D and E have lower levels of inflammatory markers, more grey matter, more positive mood states, and better visuospatial cognition than those who consume fewer of these nutrients.[122]

Conclusion

Mood is an internal, subjective state but it can often be inferred from posture and other behaviours. The more we learn about the brain and body as an interlinked system, the more it becomes clear that our perceptions of how things are is a function of deep seated memory systems that are projections of what we feel. Working to debunk lopsided perceptions and cultivating an attitude of gratitude can send us on an upward spiral of positive mood and many more good things.

6.5 Simple fluency and regulation of emotions

"If you wanted only happiness and no challenges,
You are on the wrong planet."

Anonymous

At a Glance

Controlling oneself to be socially accepted involves an awareness of emotions, including how one is thinking, feeling and behaving. In this section we look at: (1) emotional intelligence or not; (2) how emotions are made; and (3) how we can soften our grip on our emotions through deploying specific choice tactics to let our emotions work towards helping us flourish instead of languish at work.

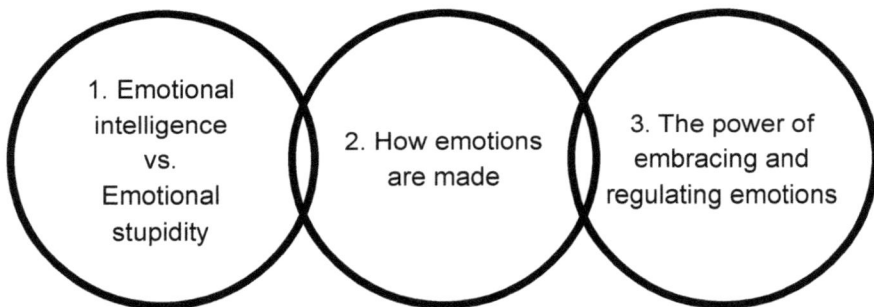

1. Emotional intelligence vs. Emotional stupidity

2. How emotions are made

3. The power of embracing and regulating emotions

1. Emotional intelligence or emotional stupidity

Emotions drive action; they are the irreducible basics of making sense of anything. If our brain is sending us the wrong emotions, then we will be engaged in the wrong actions. This links to emotional stupidity or difficulty with the concept of emotional intelligence (EQ). The whole idea of EQ presupposes that the intelligent, adult part of your brain is in control of your emotions, which it is not. To be truly emotionally intelligent, we need to understand where emotions really come from. This understanding is necessary if we want to change how we feel. If you do not know how the brain works, you will not know how to manage and improve your social and intellectual functioning. If we do not know how emotions work and where they come from, we are pretty much destined to fail in our attempts to manage or change them.

Emotions may seem to be conscious feelings, but they are, in fact, physiological responses to stimuli, designed to push us away from danger and toward a reward. Emotions can be thought of as body changes that prompt us to act. Emotions are continuously generated, but much of the time we are unaware of them. According to LeDoux[123], emotions are biological functions of the nervous system. They are not facts, therefore we should not get emotional about our emotions.

According to Rita Carter, "…emotions have evolved to get us to do what we have to in order to survive and pass our genes on to the next generation".[124] To bolster their impact and effectiveness, emotionally triggered actions are associated with pleasant or unpleasant conscious feelings. Emotions tend to be short-lived, lasting a few hours at most, but they can lead to more persistent conditions called moods.

2. How emotions are made

As humans, we are highly interdependent – what we do affects what happens to others. It is thus very useful for us to be able to read each other's emotions in order to predict what someone might do next. We also signal our own emotions to nudge others to do what we want. In essence, reading the emotions of others is a vital skill for navigating our way through life.

Another fundamental reason for why you need to know and master your emotions is that we spend only about 2% to 10% percent of our waking hours being reflective and the rest being reflexive – reacting, responding out of impulse, defaulting to the negative and thus missing the positive. In short, much of our time is spent in the reactive side of our nature, which is why most of us have to work to improve our emotional intelligence in order to succeed.[125] But are emotions hardwired or are they constructed as we go along? There are two schools of thought.

School of Thought One: Acknowledged basic emotions

It is generally assumed that our emotional states can be readily inferred from our facial movements, called emotional expressions or facial expressions. Faces do offer a rich resource of information for steering the social world; they are a crucial role player in deciding whom to love, whom to doubt, whom to help, and who is found guilty of an offense or crime.[126]

In one of the most influential (and now also highly contested) sets of experiments carried out with cultures all over the world, Paul Ekman (cited in LeDoux)[127] showed that all human beings share the same facial expressions for six basic emotions – anger, fear, joy, surprise, disgust and sadness. This is also called the common view of emotional expression. These universal emotions are almost identical across every culture. This list of acknowledged basic emotions serves as the foundation for much research on the neural basis of emotional functions in the human brain. Ekman went on to develop a taxonomy of facial expressions called "Facial Action Coding". These six emotions then split into multitude feelings and moods.[128] These are shown in Figure 17 below – the first set is human facial expressions while the second set is the augmented reality version. Can you name them?

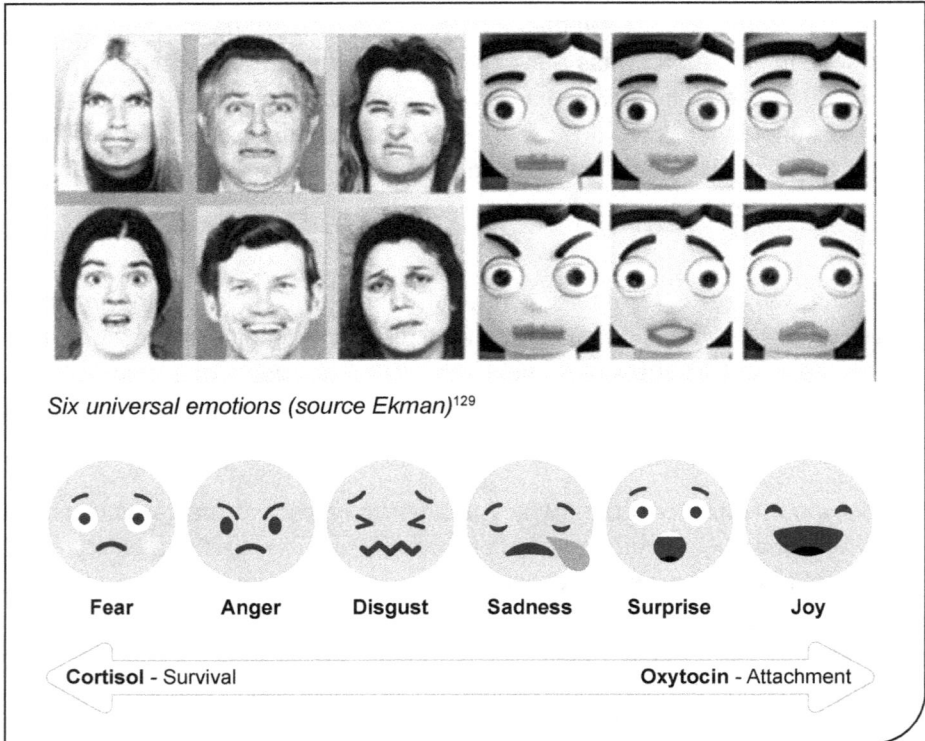

Six universal emotions (source Ekman)[129]

| Fear | Anger | Disgust | Sadness | Surprise | Joy |

Cortisol - Survival → Oxytocin - Attachment

Figure 17: Six universal emotions continuum

An atlas of emotions

Awareness of our emotions means understanding how they are triggered, what they feel like and how we respond. Awareness itself is a strategy; it helps us understand our emotional experiences so that we can respond in helpful, constructive ways. The *Atlas of Emotions* was commissioned by the Dalai Lama, who explained that, "In order to find the New World, we needed a map, and in order for us to find a calm mind, we need a map of our emotions" (see www.atlasofemotions.org). Table 7 below sets out the signal of an emotion which describes the universal ways that emotion is displayed in the face and voice.

All emotions have a message that is a response to the world around us. Sometimes responses are helpful and other times they are unhelpful.

Table 7: Signals of emotions (Source: Ekman)[130]

1. Fear	
Signal	Common signals are very wide-open eyes, horizontally stretched lips and raised, drawn together eyebrows. There may be movement away from the target. Screams may accompany intense fear. Lesser fear signals can include heavy breathing, a head position slightly backwards and away, and horizontally stretched lips accompanied by tightened neck muscles.
Message	The message of fear is "help me"; it can range from showing low-level concern to conveying panic.
2. Anger	
Signal	In the voice, anger generates a roar if not controlled. When anger is controlled, the voice may have a sharp edge that is very detectable. In the face, the signal includes glaring eyes, lowered brows and narrowed, tightened lips. When people hear or see an angry signal, they are typically hurt just by the perception of the signal and may retaliate with angry actions.
Message	The message of anger is "get out of my way". Anger can carry a message ranging from dissatisfaction to threat. Anger gives us energy to overcome obstacles.
3. Disgust	
Signal	There are three facial expressions associated with disgust. The first is sticking the tongue out as if the person is getting something out of their mouth. The second is raising the upper lip, but it is relaxed and not tense, which can display gums and teeth depending on the shape of the mouth. The third is the wrinkling of the nose and raising of the nostrils. These expressions can occur separately or in unison.
Message	The message of disgust is "get away from this". It can show others that the target of disgust is to be kept away from or that the target is unclean, dirty or socially/morally reprehensible.
4. Sadness	
Signal	The signals of sadness include a frown (lower lip pushed up slightly and lip corners pulled slightly down), the inner corners of the eyebrows drawn up and together in the centre of the forehead, raised cheeks and tears. The vocalisation of sadness can include sobs and heaving, and quivering of the voice.

Message	The message of sadness is "comfort me". It encourages, or intends to encourage, empathy from others.
5. Surprise	
Signal	Surprise and fear are two of the most commonly confused facial expressions because they are shown in the same key features: eyebrows, eyes and mouth. In surprise, the eyebrows are raised but show more curve than seen in fear. The upper eyelids and jaws are also more relaxed when expressing surprise.
Message	The message of surprise is "stop and think", which can be good or bad. The briefest emotion, surprise, is triggered by the sudden occurrence of an unexpected event. It is often a way station that leads, after more appraisal, to any of the other emotions.
6. Enjoyment	
Signal	Enjoyment signals include the Duchenne (authentic) smile, activation of a smile (lip corners pulled obliquely up), and activation of the orbital eye muscles that tighten the lower eyelid and create wrinkling around the outer eye corners (especially with age). Enjoyment also includes vocal signals such as the sound of relief (a sigh or exhalation) and the sound of amusement (laughter or giggling).
Message	The message of enjoyment is "this feels good". It encourages engaging in social interaction.

The common view (Basic or Acknowledged Emotions Theory) also claims that emotional categories are biologically hardwired via dedicated neural circuitry or their own neural substrate. For example, the amygdala, which is part of the limbic system, is implicated in detecting fearful stimuli and the insula is associated with disgust.[131]

Basic emotions like fear and anger are innate, fast, trigger behaviour with a high survival value and make us less creative. Increasingly research shows that emotions are constructed by us.

School of Thought Two: Emotions are culturally constructed

Constructed Emotion Theory states that all emotions tap into a system termed "core affect", which is organised along two dimensions: pleasant-unpleasant and high/low arousal or activation. It works as follows: all individual emotions and moods fall somewhere within this two-dimensional continuum. In biological terms, this is linked to bodily feelings of emotion and our limbic structures (see Figure 18).[132]

The unique aspect of the model is the notion that groups of emotion are created (and can be distinguished from each other) because they tap the core affect system in somewhat different ways and because they are linked to certain kinds of information processed outside of the core affect system, including executive control (for regulating and appraising emotions), language (for categorising and labelling), mentalising (for conceptualising others' emotions), and so on.[133]

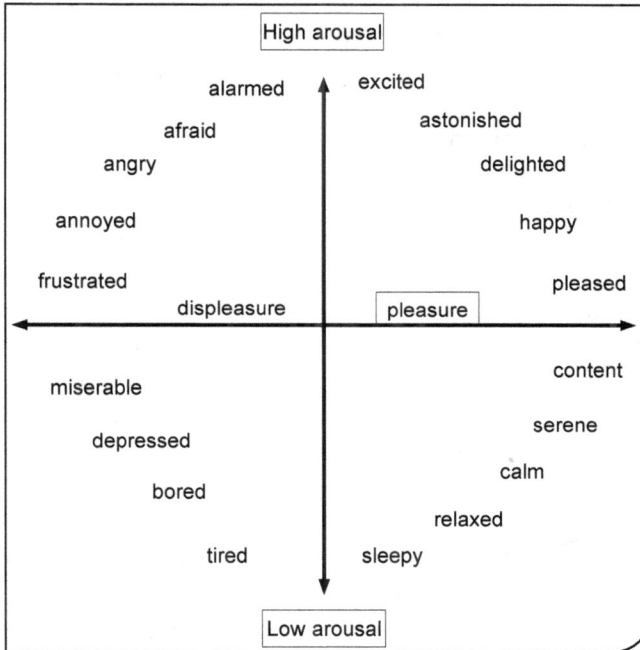

Figure 18: *Affective circumplex* (Source: Russell & Barrett)[134]

Emotions (and mood involve a 'core affect' system that is organised along two dimensions corresponding to pleasantness and activation (or arousal). Different categories of emotion are points in that space (and linked to associated cognitions – language, memory, perception, theory of mind) but are not afforded a special status.

How emotions are expressed

There are no universal expressions of emotions; the available scientific evidence advocates that we do sometimes smile when happy, frown when sad, scowl when angry etc., as proposed by the common view, more than would be expected by chance. This is not consistent, however – reliability scores are typically low and inconsistent. There is also a lack of specificity (i.e. there is no unique mapping between a pattern of facial movements and instances of an emotion category).[135] Thus we frown when we are angry, when we concentrate and when we are unsure.

The point is that we move our faces in many different ways during the same emotion category; we move our faces in situated ways which makes sense for the specific situation we are in.

Meta analyses (i.e. statistical summaries of hundreds of experiments) show that there is no reliable relationship between an emotion category (such as anger or any of the six universal emotions) and a specific set of physical changes in the ANS (autonomic nervous system) that accompany the occurrences of that category.[136] In essence, when it comes to expressions of emotions, variation is the norm.

How emotions are perceived

Identifying emotions from facial expressions is an important people skill in the workplace. How does it work? When you see emotion in another person, like a colleague, or when your colleague sees emotion in you, what is happening is that your brains are guessing (by taking in a host of data) in order to determine what those facial movements mean. These guesses occur automatically and effortlessly. This is how we make meaning of the raise of an eyebrow, the curl of a lip or the tilt of a head. This is how a single physical feature, like a smile, can take on different meanings in different situations.

The emotions that you seem to detect in other people are partly coming from you. How one views the self affects how we view and relate to others. We do not read each other's emotions or body language; physical movements of the face and body are not a language to be read. We are guessing – our brains are guessing. We often make really good, efficient guesses, but what gets us into trouble is that sometimes our guesses are not so well-tuned to the situation. Mindfulness practices can help us suspend our guesses or tone them down, but they still remain a guess.

There are huge variations across and within cultures when it comes to reading emotional expressions, and even across people within a single situation.[137] Identifying emotions generally follows along cultural alignment. We are more accurate at recognising the emotions of people from our own culture than from others, and we use the stereotypes of our own cultures to make guesses about emotions.

Due to this mental short cut, we need to take extra care to understand how emotions are processed and how we can manage our emotions. This is like reading emojis on your text messages – they need context or otherwise they can lead to painful misunderstandings.

Softening your grip on your emotions

Although we cannot control the weather outside, we are capable of using many emotion regulation strategies to take control of our internal climates. A reasonable goal for the workplace is to have some level of emotional expression, but we also need to know how to control it.

Emotion regulation forms part of a person's self-control repertoire, which, in essence, is the experience of applying effort to overcome something. The ability to put your attention where you want to, instead of on other temptations, is called cognitive control or willpower. Cognitive control is a limited resource; chronic high stress can result in one losing control and flying off the handle.

Controlling our emotions is important because it enables us to maximise our brain resources. How we respond to an emotional stimulus is the main determinant of successful emotion regulation. We differ in the ways that we experience, express and regulate our emotions, as well as in how these emotional processes affect our lives, including consequences for affect (e.g. feeling good vs. bad), for relationships and social bonds (e.g. closeness to others, relationship satisfaction), and for adjustment and psychological functioning (e.g. depression, well-being).

3. The power of embracing and regulating emotions

Research by Aldoa and Nolen-Hoeksema[138] has shown that we use a multitude of emotion regulation strategies (some even simultaneously) to help us handle our emotional responses. However, to simplify things, we use a simplified version or framework for what occurs in real life. According to Gross and John[139], emotions can be modulated or changed, and modulation is what determines the final emotional response. Emotion regulation concerns this modulation of emotion in order to alter what emotions are experienced, as well as when and how they are experienced in order to achieve a goal.

The stepwise or process model of emotional regulation (as set out in Figure 19) details five major points of focus during emotion regulation. Antecedent strategies; which refer to tactics that are implemented before emotion response tendencies have become fully activated, include situation selection, situation modification, attentional deployment and cognitive change. Response modification-focused strategies refer to things that people do once an emotion is already underway and the response tendencies (behavioural and physiological) have been generated (commonly known as suppression).

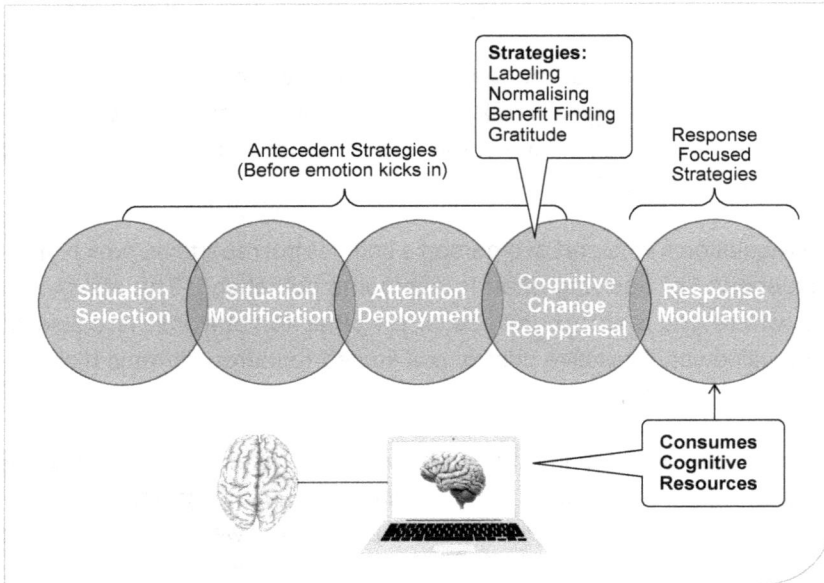

Figure 19: The stepwise model of emotional regulation choices (Source: Gross, courtesy of James Gross)[140]

Learning to regulate our emotions is not quite a fork in the road decision, but structure helps. The distinction between these strategies (or tactics) does seem blurred, however the strategies (or tactics) of affect regulation through **reappraisal** and **suppression** have **consistent findings** in the literature.

The five emotional regulation choice tactics

Choice Tactic One: Situation selection

Most of us control or adjust some of our emotions by choosing to enter certain situations and avoid others during the course of daily life. For example, you may decline a party invitation in order to prevent the feelings of embarrassment that you have come to associate with social situations, or you could skip the Eiffel Tower on your tour of Paris in order to avoid experiencing the fear that you associate with high places.

Overplayed situation selection: Situational avoidance and social withdrawal

In the case of extreme anxiety, the use of situation selection to regulate emotion and mood becomes problematic, because persistent avoidance of safe situations maintains fear, negatively affects psychosocial functioning, and diminishes quality of life.[141]

Choice Tactic Two: Situation change or modification

If complete avoidance of a situation is not possible, we often turn to situation modification tactics to regulate negative emotions. Situation modification involves making changes to a situation that alter its emotional impact. This involves tactics like timing or seeking support to reduce emotional responses.

Overplayed situation modification: Safety signals

Safety signals are things we use to reduce distress in feared situations. Examples of safety signals include medication, foods or beverages that are believed to alleviate or prevent fears and anxiousness. A cell phone is a good example that can be used if a need for help arises.

Safety signals temporarily reduce fear and afford a sense of feeling safe in feared situations, therefore they can be construed as tools for emotion regulation. At first glance, the use of safety signals may even seem like an effective form of emotional control. However, the use of safety signals is counterproductive in the long-term because security becomes associated with the talisman (or crutch) rather than the feared situation itself.[142] In other words, we never learn that the situation itself is safe because safety and success are attributed to the presence of the safety aid.

Choice Tactic Three: Attention deployment

This involves placing one's attention on a different aspect of the situation or focusing on something else to ensure attention is redirected to enable goal achievement.

Overplayed attentional deployment

Thought suppression, distraction, worry and rumination. Problematic emotion regulation strategies involve shifting attention either toward or away from the source of the negative emotion. Thought suppression involves efforts to make thoughts "go away"; it is where we deliberately shift attention from the perceived offending content (or person) onto some other target that is deemed acceptable (e.g. a "good" thought or a happy face). We now know that there is a paradoxical increase in unwanted thoughts that occurs subsequent to thought suppression, i.e. suppressing thoughts is counterproductive.

Choice Tactic Four: Cognitive change

Cognitive change is done by labeling the emotion, developing granularity, or reappraising it. This requires reason, a relaxed physiological state, and a deliberate choice.

Affect labelling – 'Name it to tame it'

The simple act of labelling an emotion serves to dampen the limbic activation (the actual feeling of the emotion) and our response. For example, before a presentation, you may label an emotion as 'nervous'. By doing this, the prefrontal cortex is activated into cognitive thinking. The braking system is activated (right ventral lateral PFC). This in itself reduces our emotional response and reflexive actions, and diminishes negative feelings.

Develop your emotional granularity or emotion differentiation

To survive in the 21st century, you need to become an Emotional Granularity Expert. What does that mean? Emotional granularity (hereafter, granularity) is the ability to experience emotions in a precise manner and is a kind of emotional complexity associated with emotional and social wellness. Granularity refers to the ability to recognise, identify and express a full range of emotions.[143]

By more clearly identifying our feelings or by re-categorising them, we can reduce emotional turmoil. Since our brain essentially constructs our emotions, we can teach it to label emotions more precisely and then use this detailed information to help us take the most appropriate actions – or none at all. It turns out that the words that we know for emotion are like tools that our brain uses to make meaning of physical sensations, as well as to predict and tailor our actions to specific situations.

As a start you need to go beyond the six universal emotions of anger, fear, joy, surprise, disgust, and sadness; add more categories to the emotions so your brain will have many more options for predicting, organising and perceiving emotions, providing you with the tools for more flexible and useful responses. Distinguish finer meanings, for example, 'angry' can represent being alarmed, spiteful, grumpy, resentful, envious etc.

People who are skilled in granular emotional experiences have a texture to the way they talk about an emotion, both about what they are feeling and the intensity with which they are feeling it.[144] They issue predictions and construct examples of emotion that are fine-tuned to fit each exact situation. At the other end of the scale are young children who have not yet established adult-like emotion notions and who use "sad" and "mad" interchangeably. Adults run the whole range from low to high emotional granularity, so a key to real emotional intelligence is to expand your emotional vocabulary so as to gain new concepts of emotions and hone your existing ones.

An emotional granularity example at work: The new project is busy derailing, but instead of saying, "I am feeling annoyed", you could add granularity by saying, "I am worried that we will not have enough time to re-scope the project". Or when you are feeling anxious before a presentation, framing it as "getting your butterflies flying in formation" changes the experience or feeling of anxiousness to that of determination.

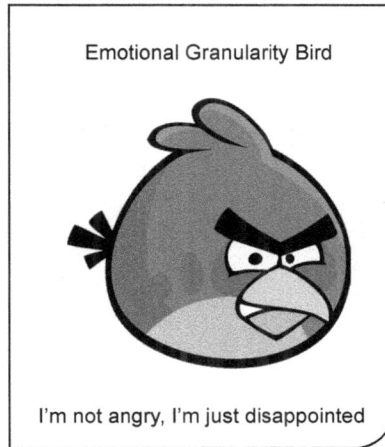

Figure 20: Emotional granularity bird (Source: Fosslien & Duffy)[145]

There is evidence that emotional granularity improves mental health.[146] Higher emotional granularity translates to better coping skills and, therefore, fewer maladaptive behaviours such as addictions. Finely grained feelings allow us to be more agile at regulating our emotions, and less likely to retaliate aggressively against someone who has hurt us. Relationships also improve when people are attuned to emotions.

> ### (?) # Neuro-Insight: How emotionally granular are you?
>
> Do you have difficulty identifying your feelings? Do you ignore them? Lump them together?
>
> Coaching, mentoring and talking therapies can help because they provide a safe place to learn about and discuss emotions. By becoming more tuned in to them, you will up your emotional intelligence and flourish in life.
>
> ### NOTES
>
> _____
>
> _____
>
> _____
>
> _____
>
> _____
>
> _____
>
> _____

Reappraisal: Changing emotional impact through realistic and evidence-based reinterpretations of emotion-provoking situations

Reappraisal has a stronger braking effect on the brain, so it reduces impulsivity to react to a situation without thinking it through first. If we can reappraise an event by reinterpreting or repositioning it towards the positive, this improves our cognitive abilities and allows us to respond more appropriately. For example, one can reappraise feeling 'afraid' of doing a presentation as being 'excited'. Anxiety (feeling afraid) and excitement are both arousal emotions and have similar physical symptoms. By reframing your stress as excitement, you can use your nerves to help you perform better. Thus, do not be afraid of your fears – they are not there to scare you. They are there to let you know that something is worth it.

Recurrent use of reappraisal has the long-term effect of generating an enhanced control of emotion and interpersonal functioning.[147]

Overplayed cognitive change = Rationalisation

This means "telling yourself" something about an emotional situation that is supposed to make you feel better. Merely thinking in a 'Pollyanna' manner (i.e. being excessively cheerful or optimistic) or justifying one's problems can actually be unhelpful forms of cognitive emotion regulation. Rationalisation ultimately rings false and may hinder one from taking action that will help solve the problem at hand, therefore it is ineffective at resolving negative emotions and moods.

Choice Tactic Five: Response modulation

Once you have felt the emotion, you can express it or suppress it. Expressing it can be damaging and inappropriate in the workplace. Since suppression has been found to reduce memory and increase the stress felt by the individual, as well as those around them, this is not ideal either. Suppression in this sense is not suppressing the emotion, but rather suppressing or trying to control one's facial expressions, tone of voice, and body language to make sure others cannot tell what you are feeling on the inside. This brute force way of emotional regulation is distressing – acting cool while feeling hot is never easy.

The neural footprint of suppression and reappraisal

Based on an analysis of how emotions unfold over time, it has been shown that reappraisal and suppression have their primary impact at different points of the emotion-generative process. Specifically, reappraisal is an antecedent-focused strategy that acts before the complete initiation of emotion response tendencies has taken place. It might thus be anticipated to modify the entire temporal course of the emotional response.

Reappraisal has a stronger emotional braking effect on the brain, with the result that it reduces impulsivity and improves our cognitive abilities. We then respond more appropriately and acceptably for the situation at hand. Frequent use of reappraisal has the long-term effect of generating an enhanced control of emotions and interpersonal functioning.[148] These two commonly used emotion regulation strategies: reappraisal (changing the way one thinks about a potentially emotion-eliciting event) and suppression (changing the way one responds behaviourally to an emotion-eliciting event) have different neurological coordinates and results. Suppression increases activity in areas involved in generating emotional responses (the amygdala and insula), diminishing the brain's capacity for higher intellectual functions. This turns out to be more physiologically arousing than expressing what is felt over time, and this can lead to long-term health problems like diabetes and cardiovascular disease for both leaders and team members.

Suppression involves hiding emotions after they have occurred. The effects of suppression are mostly adverse, not only inwardly (i.e. suppression reduces short-term memory and causes blood pressure to go up) but outwardly as well (i.e. suppressors are perceived to exhibit a lack of concern or interest in conversations, as well as a lack of responsiveness). Thus, suppressors also raise the blood pressure of others.

Reappraisal reduces activity in the limbic system, so choosing to reappraise rather than suppress can prevent a negative emotion and build emotional consistency, which is a tremendous tool for achieving leadership and business success by deploying the prefrontal cortex (the region in the brain that supports higher intellectual processes). See Table 8 below for a summary of these two regulation strategies.

Table 8: Suppressors and reappraisers

	Suppressors	**Reappraisers**
Mood:	Experiences greater negative emotions	Experiences greater positive emotions
Affect:	Expresses lesser positive emotions	Expresses greater positive emotions
Relationship:	Behavioural avoidance and worsened interpersonal functioning	Improved interpersonal functioning
Adjustment:	Negative well-being	Positive well-being

"A human mind is a wandering mind, and a wandering mind is an unhappy mind. The ability to think about what is not happening is a cognitive achievement that comes at an emotional cost."

Matthew Killingsworth[149]

Overplayed response modulation: Substance use

Response modulation is an attempt to alter the subjective, physiological or behavioural manifestation of emotion as directly as possible. Suppression is a form of response modulation that does not appear to be particularly effective for managing negative emotions, and may have undesirable consequences (e.g. increased sympathetic activation). To relieve the tension of behaviour that results in too much of everything, people tend to engage in substance use, which can become problematic and habitual.[150]

? # Neuro-Insight: A snapshot at work – Emotional fluency

You are discussing a project with a new client. The situation is difficult for two reasons. First, it is an important deal for your career and you are anxious to close it to your advantage. Second, the client is quite condescending. The pressure to succeed, and the need to refrain from getting angry and letting your emotions run wild, make it hard for you to think straight. This is because emotion and cognition are tightly interconnected. Being emotionally fluent or having the capacity to sense emotions, control them (including stress and anger) and transform what you feel into healthy action, is crucial for performing successfully in anything.

So, the lesson is: a healthy brain with well-developed capabilities is essential for all aspects of life. Ultimately, the human brain has evolved to help us operate in complex, changing environments by continually learning and adapting. Successfully doing so involves a variety of interdependent brain functions and abilities.

Cognition has to do with how one understands and acts in the world, and cognitive skills are the brain-based processing capacities we need to carry out any task, from the simplest to the most complex. They have to do with the mechanisms of how we learn, remember, problem-solve and pay attention. Any task can be broken down into a set of different cognitive skills or functions needed to complete that task successfully.

Emotions are complex states that involve both physiological or bodily experience and psychological or cognitive experience. They are closely related to motivation as they often precede our actions (e.g. if we are angry, we may start arguing) or follow them (e.g. we may feel happy after helping someone). Feelings are part of the emotional experience; they are the way we consciously perceive and describe emotions.

Emotion and cognition are both essential parts of ordinary functioning. Emotion is the system that tells us how important something is; attention focuses us on the crucial things and away from those that are unimportant. Cognition tells us what to do about it; cognitive skills are whatever it takes to do those things. See http//:www.sharpbrains.com to learn more.

NOTES
_____
_____
_____
_____
_____
_____
_____
_____

"You can't manage what you have not mastered."

Unknown

Neuro-Insight: Benefit finding

Neuroscience research shows that **Mindfulness and Positive Reappraisal** are the most beneficial emotional regulation strategies, as these dampen activation in the limbic system and increase serotonin in the brain. This positive reappraisal strategy entails reframing stressful events as beneficial, meaningful and contributing to one's personal values. It involves disengaging from the negative appraisal via a process of mindfulness which leads to a state of mindful attention awareness. This may result in dispositional mindfulness with a heightened propensity to make a positive reappraisal of distress in future, and lead to flourishing at work and in life (Figure 18) – an upward spiral of flourishing over time.

Research shows that individuals who rate highly on mindfulness scales also rate highly on emotional intelligence levels. The reason for this correlation is that one of the things that mindfulness allows us to do is to suspend emotional guessing or predicting emotions, or at least to tone it down substantially. The idea with mindfulness is that you do not process sensory data and do not make meaning of the sensory data, such as anger, sadness, fear or any kind of formed perception.

Mindfulness is an important skill to have – the more you practice it deliberately, the more practiced you get at it and then the more automatically you can use the skill of suspending judgement. This enables you to deconstruct your experience and to more easily take the same sensory data and reconstruct (re-appraise) it into a different experience, just by changing the meaning of the sensory data.

Different emotional regulation strategies like neutralising and benefit-finding give the brain a perceived sense of control, which over time can lead to an upward spiral of flourishing. This perceived sense of control activates the pre-frontal cortex into cognitive thinking and decreases limbic activation (emotional responses).

Emotional regulation strategies that strengthen downward spirals or attention narrowing include ruminating, catastrophising, self blame, and other blame.[151]

NOTES

_____

_____

_____

_____

_____

_____

_____

_____

_____

_____

_____

_____

Upward Spiral of Flourishing

Positive Emotions &
Decreased Stress

Positive Reappraisal

State of Mindfulness

Attentional Broadening

New Stress Appraisal

Decentering

Positive Reappraisal

Positive Emotions & Decreased Stress

Attentional Broadening

State of Mindfulness

Decentering

Stress Appraisal

Trait Mindfulness & Positive Dispositionality

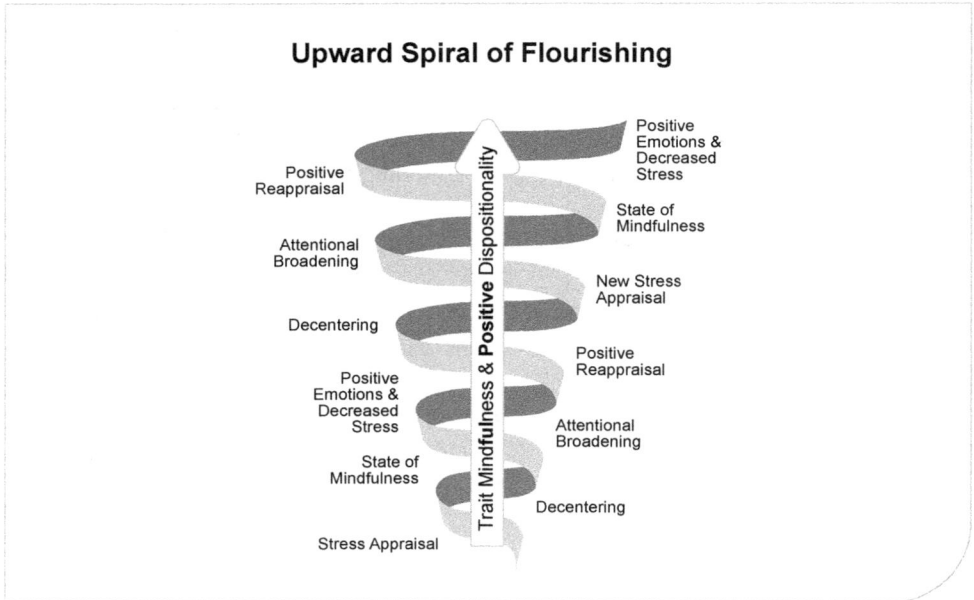

Figure 21: Positive reappraisal – benefit-finding in conjunction with mindfulness (Source: Garland et al.)[152]

Defining how having a difficult boss is assisting you to build perseverance, spend time improving yourself, learn about conflict management, and appreciate the supportive family you have is an example of positive reappraisal. Building an upward spiral of flourishing through mindfulness and positive reappraisal is a deliberate strategy; it is not 'spontaneous', but an active pursuit of positive events that drive the spiral upwards.

Neuro-Insight: Benefit finding

Whenever you have a fantasy about how you think life should be vs. being grateful about how it is, you polarise yourself. The fantasy has a ratio of more positives than negatives and as a result, in comparison, reality becomes the polar opposite – more negatives than positives. This leads to depression or feeling down. In order to dissolve feelings of depression, you simply step into the fantasy and discover the drawbacks of IF what you are imagining did, in fact, occur. As soon as you balance the ratio of positives and negatives in the fantasy, you do the same in reality and the depression dissolves.

1. What specifically are you frustrated about?

2. What are you comparing your reality to?

3. What are the drawbacks of the ideal you are comparing your life to?

4. Write down the benefits of your life as it is right now.

NOTES

_____

_____

_____

_____

_____

_____

Benefit finding is another term for gratitude. When we practice feeling grateful, we notice more things to be grateful for. This means an alteration in glial (i.e. non-neuronal) activation to fortify the new pathways – neurons that fire together, (ultimately) wire together. Over time this becomes a confirmation bias. The more I am grateful for things in my life, the more I see things to be grateful for.

"You don't need dopamine fasting in order to break your addiction to happiness. You need to embrace suffering or the dark side of your life. Just as you need a good night's sleep in order to enjoy a happy and productive day, you need to accept and appreciate the dark night of the soul in order to flourish."

Dr Paul T . Wong[153]

#Neuro-hacks for leveraging self-regulation and fluency of emotions

• Build your capacity to **identify emotions** in yourself and others (such as fear and happiness). Develop your emotional vocabulary.

• Foster an environment where people **feel comfortable expressing their emotions**.

- Develop your **pre-appraisal capacities**. Notice windows of opportunity to self-regulate and maintain mind-*flow*-ness. Integrate meditation tools and practices into day-to-day living.

- **Emotionally proofread** your messages, emails, tweets and posts before hitting the 'Send' key.

- **Thoughtfully "call out" emotional suppression** to address and improve the emotional environment. Explain the impact on others. This is also known as naming the elephant in the room.

- Ask what thoughts or beliefs are driving your emotions. **Label emotions in just a word or two**.

- Encourage **labelling and reframing of emotions** to reduce stress. For example, "I feel frustrated. What are new, more positive ways to look at this?"

- Use email and text messages (or any social media posting) like it will one day be used as a formal written statement in a High Court. Also, add your favourite emojis, but **proceed with caution as they can trigger an emotional storm and ruin a relationship or a career.**

Conclusion

Understanding the neural basis of emotions has implications for organisations, as emotions change our thinking patterns. We do not build productive relationships at work if we show up like robots; instead, we need to be fluent with our emotions. Emotions are not facts – they can be harnessed to become more creative, collaborative and productive. Learning to harness emotions, and building regulation strategies to slow down emotional reactions, are vital to succeed at work. A key take-away is to develop your emotional vocabulary

6.6 Social Connectivity

"According to most studies, people's number one fear is public speaking. Death is number two. Does that sound right? This means that to the average person, if you go to a funeral, you're better off in the casket than doing the eulogy."

Jerry Seinfeld[154]

At a Glance

Our social connectedness is at the heart of human evolution and flourishing. The depth and quality of our relationships (connectedness) are dependent on the extent of our alignment with our conscious and nonconscious processes, as well as our mutual understanding of them. The focus here is on: (1) on the building blocks of human social bonding and belonging; (2) how we copy and mindread each other to fit into the group; and (3) how sustained collaboration in groups can make us stronger together.

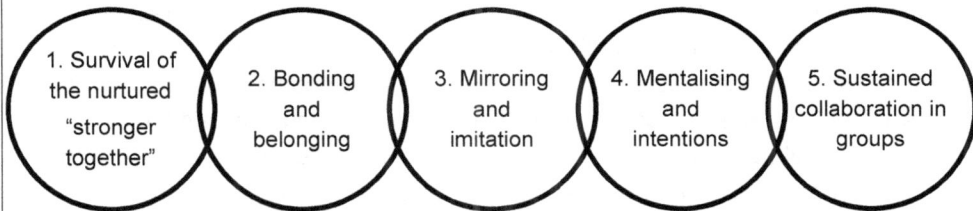

1. Survival of the nurtured "stronger together" | 2. Bonding and belonging | 3. Mirroring and imitation | 4. Mentalising and intentions | 5. Sustained collaboration in groups

1. Survival of the nurtured

Today the top skills that are sought in the workplace are the ability to work in a team and to connect and communicate with stakeholders. In this regard, neuroscience is clear – we are social first.[155, 156]

Social neuroscience is very helpful in understanding our behaviour at work. A cornerstone of social neuroscience is the social intelligence hypothesis, which argues that the evolutionary pressure for survival demanded the development of social intelligence to outwit our peers. Social intelligence – or the ability to understand and predict complex social interactions – has been correlated with larger brain size development. On this topic, the most prominent work is that of anthropologist Robert Dunbar – known as "Dunbar's Number". Dunbar[157] used social group size of various primates as an approximation of social complexity and found a correlation with neocortex ratio. The effect is that the larger the brain, the higher the number of social relationships that can be sustained. The theory of Dunbar's Number posits that 150 is the number of individuals with whom the average person can maintain stable, primary, I-care-you-care relationships.

This is simply because we reach our cognitive limits, and because there is not enough time for more relationships without spreading the quality of all relationships too thinly. Beyond the hypothetical constant of 150 relationships, we see others as objects not worthy of any care. So despite having lots of friends on social media, our brains are calibrated to handle about 150 overall relationships, and only a few close relationships.

"So what?", you may ask. Well, it turns out that, if taken as a valid premise, we can draw a few conclusions relevant to our world at work. One repercussion of Dunbar's Number is that when we change from primary to rule-mediated relationships, mutual care is substituted by structural suspicion. This shift is not trivial. By necessity, a boss, manager or leader will tend to put systems and processes before care or service. These boundary structures become impersonal, and that impersonality amasses power to itself over and against the caring for the individual.

Another inference is to treat anyone in your team as part of your 150 'I-care-you-care relationships'; take them for lunch and nurture the relationship. One of the reasons many of us feel disconnected in the face of corporate committees and boardroom crises is that we are cut off from the social support and cohesion that can only happen in small, familiar groups. So, do not substitute secondary (weak) bonds for primary (strong) ones. We are wired for strong primary bonds.[158]

Our default is social

Large-Scale Brain Network Theory[159, 160] tells us that there are multiple social networks in our brains that enable surviving and thriving socially. These networks represent key social adaptations to the environment and help us to effectively coordinate our lives with each other. These social networks also help to hold back our selfish impulses for the greater good of humanity. On the flip side, these social networks support envy and unhealthy competition at work.

The social network is called the default network (DNW) because social thinking is, in essence, a reflex. In an experiment conducted by Spunt, Meyer and Lieberman[161], the DNW was activated whenever a participant took a break from a maths exercise sample. The message is that we use our free time to think socially and a reason could be that this default 'go to behaviour' has been an evolutionary selection to adaptively ensure the survival of the species. For example, while presenting the annual budget, you become aware, amidst all the complicated financial debates and explanations, that the boss is not making eye contact. You then infer that your team's budget is not going be approved and that the team is possibly going to be reduced, which might lead to some team members having to leave.

Matthew Lieberman, one of the foremost authorities in the world on the study of social neuroscience, included a schematic picture of key social adaptations across human history in his book, *Social: Why our brains are wired to connect.* (Thank you to Prof Lieberman for the permission granted.)

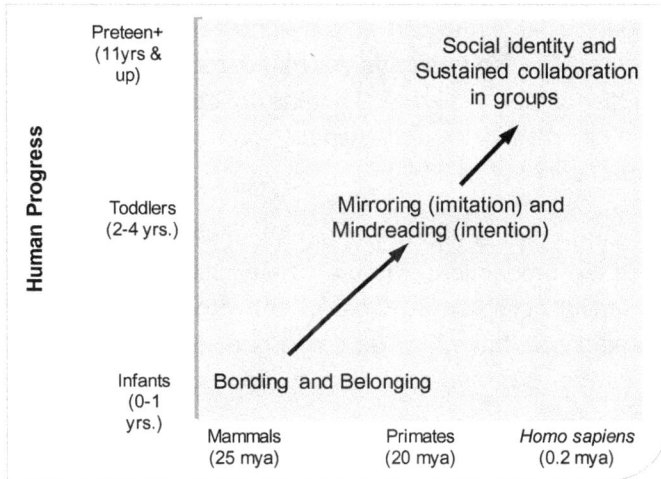

Figure 22: Our social adaptation across evolution and human development (mya = millions of years ago) (Source: Adapted from Lieberman)[162]

Bonding and belonging (connection): The earliest adaptation was the ability to feel and experience social pains and pleasures, forever linking our well-being to our social connectedness. This evolved when mammals separated from other vertebrates, long before there were any primates with a neocortex. This deep-seated need to bond and stay connected, a need best embodied by infants, remains with us for our entire lives.[163] This is a physical and emotional need for belonging. For example, it is nice to get many happy emojis in your WhatsApp messages, but a real hug feels so much better!

Mirroring (imitation) and Mindreading (intention): Being able to understand the actions and intentions of those around them, thereby enhancing their ability to remain connected and interact strategically, was an unparalleled development among primates. Especially in the infancy and formative years, the development of higher forms of social thinking, which transcend those observed among adults of other species, results in the capacity for humans to anticipate the needs and likes of others and to create smoothly running groups, capable of implementing almost any idea.[164]

Sustained collaboration in groups (synchronising and harmonising): The self, often deemed to be a mechanism to distinguish ourselves from others and possibly accentuating our selfishness, actually operates as a powerful force for social cohesiveness. This is particularly apparent during the preteen and teenage years where adolescents, in the process of focusing on their selves, become highly socialised and influenced by the beliefs and values by those around them – a process of neural adaptation defined as harmonisation.[165]

Understanding our social wiring can help us untease human workplace behaviour at its best and worst. The 'source code' for neuroscientists – the epic and immense work of Erik Kandel, James Schwartz, Thomas Jessel, Steven Siegelbaum and A.J. Hudspeth, *The Principles of Neural Science*, can be summarised in four words: *Principles of neural connection*.[166]

"Everything in neural science – from our genetic makeup, to the activity of neurons and glial cells, to the production of neural chemicals – points to one key function: managing and maximizing connections for the organism to (first and foremost) survive and ultimately to thrive."[167] The same goes for social connections at work. This requires a deeper dive into social safety, bonding and belonging, mindreading and us/them-ing.

2. Bonding and belonging

The picture below must be one of the best pictures that capture the feeling of being left out. It projects pretty much the same feeling as when you are left out of an important meeting or project at work.

Figure 23: Being left out (Source: Millner, 2013)[168]

The displeasure (or pain) of rejection

This picture talks to how our brains evolved to experience threats to our social safety in much the same way that they experience physical pain. The regions that code for physical distress also code for social rejection. Our social pain circuitry helps ensure the survival of our children by helping to keep them in close proximity to their parents.

Social pain, such as being rejected, isolated or left out from a meeting at work, produces similar and equally strong brain responses as physical pain (engaging the Dorsal Anterior Cingulate Cortex [dACC]). The Anterior Cingulate (ACC) is a neural adaptation that separates us from reptilians. Humans have a cingulate cortex, reptiles do not. The ACC also has the highest level of opiate receptors of any region in the brain. Opiates are the brain's natural painkillers; their production and release diminish the experience of pain. Social separation is wired to be painful in all mammal infants and leads to a reduction in opiates. Reconnecting with a caregiver leads to killing the pain of separation. This can even be felt as an addiction (the bliss of a mother child reunion, or re-joining a team that you felt you belonged to at work).

The ACC also plays a role in cognitive control functions as it monitors conflict and detects errors. The ACC is the key node of the brain's salient or control network and is known as the brain's error detection system. Not connecting socially is recorded as an error in the brain and the need to connect will be satisfied either in an adaptive way (approaching others to socialise) or maladaptive way (avoiding others to minimise the pain of socially painful encounters).[169]

This makes sense when trying to understand the survival of a species, but imagining the survival circuitry at work or in the boardroom almost seems too 'hot to handle', although that is, in essence, who we are. We have a deep need to be connected, which is present throughout our lives. A sense of belonging is not the same as feeling similar to everyone else. Indeed, our longing to fit in often constrains us to hide who we really are. Belonging is when we feel safe and appreciated for embracing what makes us different. We feel a sense of belonging when we are confident the team did not accept our idea because it was not the best choice, and not because something is innately wrong with us

Lieberman calls this fear of social rejection the, "Broken hearts equal broken bones phenomena".[170] Thus, pain is pain in the brain and understanding how the brain is wired enables us to use our pain as information for personal transformation and growth.

Our sensitivity to be included can even switch off our ability to think clearly. Individuals who are highly sensitive to rejection may have difficulty accessing the prefrontal cortex (PFC) regions which provide conscious cognitive control of social behaviour.[171] Extrapolating this to team dynamics and team performance, team members who manifest high sensitivity to rejection may have greater liability in social situations and therefore a higher level of defences that distort reality and enhance their coping. Should the worker with high rejection sensitivity be the CEO, it can lead to an organisational culture that is defensive and low in trust.[172] To protect ourselves against the pain of rejection, we deploy defences.

Pain killers for social pain at work: Defence mechanisms

Neuroscience has shown that social order comes from the way we are; we avoid uncertainty and strive towards comfort, even in our work lives.[173, 174] We typify and categorise or stereotype people due to constant transference from past experience, remembering past people and events and placing current phenomena into comfortable categories.[175]

Given that the family is the original and most fundamental social group, its basic structure carries over to other social groups, like work teams.[176, 177] Our social behaviour at work can only fully be appreciated by understanding our basic human needs, which are largely captured in our early attachments to our caregivers.

This nature-nurture heritage becomes a mind-set or lens through which all other relationships are filtered. If we feel unsafe at work (a fight to survive), we develop defence mechanisms or ways of protecting our sense of self when we feel under attack.[178] This is also seen as ways in which neural networks have adapted to cope with emotional stress.[179] These mechanisms range from the primitive to the sophisticated and allow team members to remain relatively functional, even when they are suffering.

Although defences are often invisible to the owner because they are organised by hidden layers of neural processing[180] that are inaccessible to conscious processing, the level of defences (or pain) in the team can be determined by watching out for:

•   primitive or misplaced defences like denial, rationalisation and projection;

•   lower level defences like passive-aggressive behaviour, regression and acting out; and

•   higher level defences such as humour or overcompensation.[181]

Higher level defences, like humour, allow team members to keep in contact with others and remain attuned to a shared social reality, whereas primitive defences, like denial, cause greater distortion and can cause difficulties in team functioning.

Defences provide a temporary relief from suffering and discomfort, like physical pain killers, but if the underlying cause is left untreated, the pain will return. In essence, feeling socially rejected is associated with greater reports of social disconnection in day-to-day life. Defence mechanisms are our strategy of protecting ourselves from the pain of social rejection. Most of us lie, spin stories or exaggerate a few times a day. Indeed, the brain will make astonishing adaptations to relieve the anxiety and other social pains related to social exclusion (real or perceived). We need to allow each other our defences as we are all trying the best we can.

Delicious inclusion

Just like social pain being similarly coded to physical pain, social pleasures are coded to the same brain circuitry as physical rewards (see Figure 24). Thus, social experience draws upon the same brain networks to maximise reward.[182]

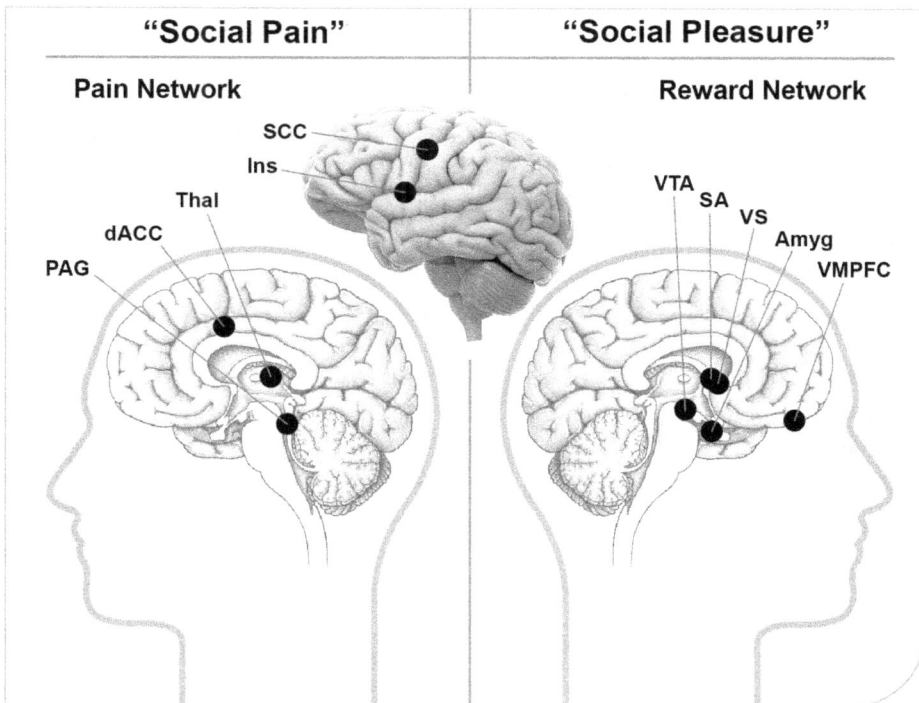

Figure 24: Social pain vs. physical pain (Source: Adapted from Lieberman & Eisenberger)[183]

It turns out that the regions of the brain that make us enjoy physical pleasures like the taste of honey (or chocolate) are also the regions that record our social connectedness as rewarding. That great feeling when you are invited to join a task team that you love, or getting recognition at the annual rewards function, equates to the sweet taste of chocolate!

The brain's reward centres include the ventral striatum and ventromedial PFC. Dopamine is released in these regions when you are being included in a group that you like, being respected by your peers, being treated fairly compared to others, and receiving positive feedback. Fairness is one of many cues indicating that we are socially connected; fair treatment implies that others value us.

Surprisingly, neuroscience research shows that helping others leads to the release of oxytocin in the ventral striatum and ventromedial PFC. So, there are two kinds of social rewards, namely the social rewards we receive when others let us know they like and respect us, and the social rewards we get when we show pro-social behaviour (i.e. 'giving is its own reward').

Care to dare

Connecting socially also goes with a sense of feeling calm, peaceful and being nurtured. This feeling comes from a combination of opioids and oxytocin. When we form a social bond, oxytocin reconfigures the opioid system so that being in the presence of the social bond (in whichever shape and size), stress and pain is relieved. In the absence of the person, team or working group, distress increases. A great analogy is that 'being cared for' (a strong social bond) is like a nurturing thick soup, while the feeling of not being cared for equates to the compromised taste of a thin watery soup. Only once we are cared for, can we venture out and dare.

Being cared for promotes opioid-based pleasure processes in the brain. Oxytocin modifies dopaminergic processes to promote approach behaviour in the bonded relationship. In humans, it has both caring (including others) and aggression (isolating others) effects, but sometimes it is a bitter-sweet symphony with a narrow line between cooperation and competition.

The narrow line between love (cooperation) and jealousy (competition)

Social neuroscience shows that as humans we divide others into at least three categories to decide if they are friend or foe: members of liked groups (positive associations); members of disliked groups (negative associations); and strangers whose group affiliations are unknown.[184] To untease this a bit, we need to look at the not so clear relationship that comes with the title of "frenemy". In the workplace, the frenemy dynamic is, for example, characterised by competition with, and envy towards, our own team members, or other teams.

The green-eyed monster – envy and jealousy

"Comparing is a trap that permeates our lives, especially if we're high-need-for-achievement professionals."

Thomas DeLong[185]

The neural correlates of social comparison emotions like envy, jealousy and resentment were mapped by researchers who showed that when the target person's possession was superior and self-relevant, stronger envy and stronger ACC activation were induced. The brain reads it as, "Something is wrong here!"[186] Stronger striatum (reward) activation was induced when misfortunes happened to envied persons. So, the green-eyed monster has a place in the social world, and it is here to stay. There is even a handbook about jealousy entitled, *Jealousy: Theory, research, and multidisciplinary approaches.*[187]

The evolutionary underpinnings of this research on jealousy talks to how we are hard-wired to experience another's failure as rewarding, also known as 'schadenfreude', as it subconsciously reduces competition from siblings (team members) and enhances the probability of approval by the parental figure (the boss). Thus, when we perceive that our existing social bonds are threatened by the perceived interventions of a third party, our basic emotional substrate of jealousy kicks in. Put another way, we tend to choose friends who are like us, especially at work, but the more we have in common with someone, the more likely we are to compare ourselves to them. The technical term for this is "ambivalent relationships", i.e. the frenemy.

Frenemies are both our friends and our benchmarks within the organisation, and on average about half of our social network consists of people we both love and hate.[188] Ambivalence has many benefits in a way – it feels uncomfortable and discomfort arouses our brain, which leads to better performance. Despite some benefits, it is not a pleasure to have most of our relationships be ambivalent. There are many more upsides to having positive relationships and that is where our focus should be.

On a practical level, relax – the occasional green-eyed monster is part of life; you are not alone. Jealousy/envy is learned behaviour and can possibly be unlearned by providing a socially safe and fair workplace where social bonds are deliberately built. Also, understanding the neurochemistry of jealousy can be helpful in coaching, counselling and team development, where these learned perceptions can be dissolved. Mindfulness practices also assist in processing emotional defences and act like vitamins for the soul.

The neurochemistry of social sentiments

The Reassuring Molecule, the hormone oxytocin or the 'bonding hormone', which affects behaviours such as trust, empathy and generosity, also affects opposite behaviours, such as jealousy and gloating. Oxytocin is an overall trigger for social sentiments; when a person's association is positive, oxytocin bolsters pro-social behaviours. When the association is negative, the hormone increases negative sentiments – in essence, a double-edged sword.

Oxytocin (a little molecule) has been linked to many positive words. To organisations it can offer reassurance because oxytocin influences social behaviour. It can also reduce stress reactivity, increase our tolerance for pain, reduce distractibility, and help retain executive control. Paul Zak[189] suggested that oxytocin is produced in high-performing workplaces and that if you want to keep people on task positive behaviour most of the time, you want oxytocin-producing situations.

One way that oxytocin levels are raised in organisations is through trust. It is suggested that trust is a temporary attachment between people. We know that oxytocin levels rise with a social signal of trust and oxytocin is associated with trustworthy behaviour (when people reciprocate trust).

Yet oxytocin can also enhance our generosity towards complete strangers in what is called pro-social behaviour when, for example, we support/give to a cause or charity. This is supported by engagement studies that show that companies that have active social responsibility projects where employees are involved in self-transcending actions results in higher commitment of employees to the organisation.

? # Neuro-Insight: ALWAYS inwardly investigate

Whenever you feel that someone has treated you unfairly, you have the option to perceive yourself as the victim of your history and 'outwardly retaliate', or to master your destiny and 'inwardly investigate'. This does not mean that there may not be a place for outer retaliation at times. It just rarely results in a clear and poised mind afterwards, unless true and fair resolution has been negotiated and eventually emerges.

Revenge is an incomplete awareness. It is wiser to define clearly what you perceive the other person or group has done and what you feel is unfair. Then it is beneficial to identify where and when have you done this action in your own unique way. It is unwise to judge another for what we are doing or have done ourselves.[190]

NOTES

_____

_____

_____

_____

_____

_____

_____

#Neuro-hacks for leveraging social belonging

- Cultivate the **'tend and befriend response'**. This is about reaching out and asking for help; connecting with someone, talking, asking, sharing, helping a colleague. If we can shift our response to connecting with others we can truly change the world.

- **Pain (social rejection) is information**. Learning to use your pain for growth and transformation is black-belt personal development stuff. The truth is that duality is a universal principle; there cannot be day without night, positive without negative, an over-performer without an under-performer, social inclusion without social rejection. So, you will be both rejected and included at work. Think of it, rejection has always protected you for something greater. Next time you feel rejected, do not take it personally. Know that you are being protected for something greater so that you can grow.

- **Build trust in the workplace**. Earning the trust of one's colleagues is not just a soft, nice-to-have asset – it is hard currency that can mean the difference between success and failure. Social trust is an essential element for the facilitation of social interactions, including the ability of leaders to trust their followers. This, in turn, promotes openness, transparency and truthfulness. The neurochemistry underpinning social trust is mainly the peptide oxytocin found in the brain, which has been found to be a chemical antecedent to trust. When this simple molecular hormone is diminished as a result of perceived rejection, or perceived being ignored, the ability to trust declines. When it is increased, it down-regulates the release of cortisol.

- **Work towards 'full disclosure' in your interpersonal transactions**. This is not easy but possible when you learn to down-regulate/dial-down fear, uncertainty and being right. Learn to up-regulate/dial-up transparency, understanding, empathy and other meta-skills that enable a safe environment where team members can communicate and reflect in a real and honest way and can fully disclose their fear factors (see Figure 25 below).

	Down-regulate	Up-regulate	
F	Fear	Transparency	**T**
E	Ego-trips or being right	Relationship building	**R**
A	Avoidance to act	Understanding	**U**
R	Revenge and resentments	Shared vision of success	**S**
S	Self-righteousness and self-wrongness	Truth and empathy	**T**

*Figure 25: Down-regulate/Up-regulate (*Source: Adapted from Glaser)[191]

There are various applied organisational neuroscience models to use that can help build trust and finally flourishing at work. In essence these are classifications of what motivates/demotivates us socially. The common themes across these models are providing **Clarity, Choice, Collaboration, and Consistency (fairness) in the workplace**. See Appendix A for a deeper dive into these valuable models.

• **Know that trust is not fail-safe**. Our brains make a determination of the trustworthiness of everybody within milliseconds of meeting them. This initial estimation continues to be updated as more information is gathered and processed. The brain is simultaneously appraising physical appearance, gestures, voice tone, the content of spoken communication, and many other factors. This happens so quickly that most of us will find it difficult to express exactly why we trust or distrust a person.[192]

• **Engage your executive function to neutralise emotional charges of distrust**. Trust lives in the PFC, and engaging this part of our brain neutralises perceived threats and enables us to see other options of partnering and co-creating something good.

• **Reframe the role of your "ambivalent relationships" (perceived frenemies)**. Frenemies provide a basis of motivation and working alongside them will make you work harder to prove yourself. The time you spend in each other's company will also help you understand each other better and perhaps even develop some empathy. Even if these relationships do not make it to the "friend" circle, they have some unsuspected benefits.

Conclusion

Our social connectedness is at the heart of human evolution. Threatening our social connection is painful. Caring for and nurturing our social bonds is highly rewarding for both the carer and the cared for. The neural connection between social and physical pain ensures that staying socially connected will be a lifelong need, like the need for food, warmth and shelter. Our sociality is our default. Knowing this, we can learn to be our best selves and to use social connections to build more cohesive groups and organisations.

Imagine

A workplace where trust and respect are actions, not just words

3.   Social awareness – mirroring and imitation

*"Mirror neurons show how strong and deeply rooted is the
bond that ties us to others."*

Giacomo Rizzolatti and Corrado Sinigaglia[193]

Social chameleons – the what and how of behaviour

People in relationships become more emotionally similar over time. This similarity or convergence helps coordinate the thoughts and behaviours of the partners, increases their mutual understanding, and fosters their social cohesion. In essence, we imitate each other to enhance conversational flow and understanding. These emotional processes and their co-ordination across interaction partners are of central importance to relationship formation, functioning and long-term outcomes. This relates to the chameleon effect – we are social creatures and we have a tendency to imitate one another automatically. Thus, we are all copycats. Social cognition has come to encompass a broad range of mental processes, but in the strictest sense, social cognition is about understanding self and others. How does this happen? The neurobiology of this imitation process can be explained through the mirror neuron system[194, 195], which allows us to think about others, and specifically, to imitate others (the 'what' and 'how' of their behaviour).

Mirror neurons fire both when we are carrying out a particular action and when we observe that action. The brain circuits for thinking about self are activated in a similar way when thinking about others.[196] Thus, at the most basic neural levels, we are modelling others and learning from their behaviour by directly simulating them in our minds.

Mirror neurons function in the same way for sound as they do for observation. The implication for the workplace is that organisational language matters and can have a large influence on work teams and performance. The mirror neuron system responds differently, depending on whether we are observing someone who shares our culture or someone who does not. Culture and ethnicity influence the brain activity involved in social communication and interaction. The implication is that we interpret cultural signals and align our behaviour to the norm – this can contribute to the experience of us-vs-them. The mirror neuron system is also activated when we learn a new skill. Observation directly improves muscle performance via the mirror neuron system (MNS), and our ability to perform an action is considerably improved when we have watched someone performing that action.

The leader becomes the organisation

When it comes to diffusing emotions, some individuals have more influence in passing along their feelings. When there are power differences between people, the person with the most influence is the 'sender' of feelings. Ralph Waldo Emerson observed that, "Organisations are the lengthened shadows of their leaders".[197]

Thus, relationship partners with less power (i.e. team members) make more of the change necessary for convergence to occur. This convergence becomes contagious and can have a lasting impact on individuals and groups. The brain's mirror neuron system also underpins the construct of emotional contagion. The existing evidence shows that while all emotions can be contagious, negative emotions have greater power to influence. This can lead to a negative organisational culture.

On the positive side, mirror neurons result in followers detecting the leader's smiles and laughter, prompting smiles and laughter in return.[198, 199] On the negative side, a leader who is self-controlled and humourless will engage those mirror neurons in his/her team members, leading to team culture that is self-controlled and serious. On the extreme end, even a bullying culture can be created that can permeate the organisation (should the bully be the CEO).[200]b Leaders who can self-regulate in favour of positive affect and a good mood will help team members take in information effectively and respond openly.[201] The basic objective is thus for neurally-aware leaders to create a safe environment for all individuals within the organisation. The effects of activating neural circuitry in followers' brains can be very powerful.

#Neuro-hacks for leveraging the Mirror Neuron System

- Take accountability for the impact you have on your work team as you might unintentionally transfer your own feelings of distrust and negativity to others.

- Be aware of the domino effect; if you are upset or emotionally unstable, you can set up a chain reaction that distracts others from optimal performance.

- Organisational language (tone and content) matters and can have a huge impact on work teams and performance.

- Do self-monitoring, as the essence of the mirror neuron footprint of emotional contagion means that you copy powerful others (for better or for worse).

- Actively monitor yourself for contagion of emotions, specifically negative emotional contagion; use positivity/negativity ratios to keep track of your mind-state.

A word of caution – while mirror neurons' activity correlates with attempts to understand other people's actions, their involvement seems neither necessary nor sufficient and is mostly associated with low-level, concrete aspects of understanding others (like what and how others are doing things). Thus, the punchline on mirror neurons and their link to empathy is that there is a correlation but not causation, i.e. it has not been proven that mirror neurons mediate empathy.[202]

4. Understanding others: mentalising and intentions

"Knowledge is Power but Understanding is everything."

Unknown

Within the leadership literature, social awareness is about one's ability to understand others. Understanding how others' minds work can help us get along. In addition, our ability to guesstimate the what, why and how of others' thinking helps us to collaborate with others in groups.

Clearly, our brains respond to others in ways that are involuntary and spontaneous. As such, our responses are often reflexive. Peering into the minds of others to understand their intentions and the meaning behind their actions or inactions is called mindreading, and is built on the premise that we have a theory about how others' minds work. The ability to identify "what" someone is doing is the first step towards being able to understand "why". The mirror neuron system provides the premise that the mentalising system can then logically operate to answer the "why" question.

When we imagine or are thinking about the intentional mental states of others (for example, their feelings, beliefs, needs), we use our own mental state as a reference. Thus, mentalising about others is self-referential. Consequently, our ability to understand and think about others, and even our entire team's dynamics, are directly interconnected to our self-understanding. Accordingly, if you carry a lot of anxiety and unresolved conflict, you will project that onto the team or 'read' the team in the same way, and you will not be able to create an environment where the team feels safe to drop their defences.

Mentalising or mindreading

The terms "mentalising" or "mindreading" refer to the process by which we make interpretations about mental states. Thus we have a theory of mind (ToM) which enables us to distinguish between our self and others.[203] This is a highly complex function and most of the time these inferences are made automatically, without any thought or deliberation. Mentalising, in particular, refers to our ability to read the mental states of others and involves several neural processes. The brain's mirror system has a role to play in sharing the emotions of others.

The human brain has the unique ability to represent the mental states of the self and the other, as well as the relationship between these mental states, making the communication of ideas possible. It is important for us to be able to read the minds of others because it is our mental states that determine our actions. This assumption that behaviour is caused by mental states has been called "the intentional stance"[204] or "having a theory of mind".[205]

There are many types of mental states that can affect the way we interact with others. There are long-term dispositions: one person may be trustworthy and reliable while another is hopelessly volatile. There are also short-term emotional states like happiness and anger. There are desires like thirst and their associated goal-directed intentions (e.g. getting a cappuccino from the coffee bar). There are the beliefs that we have about the world which determine our behaviour, even when they are false (someone has secretly removed the doughnuts from the cupboard). Karl Jung[206] introduced the phrase "perception is projection", meaning we see attributes in others that we possess ourselves.

Because mentalising about team members results in minimal output, it is tempting to try to rush through or short circuit this mental activity, instead hoping that the group can achieve peak performance through task orientation only. Mind reading is not perfect – we do make mistakes, for example making generalisations, group think, false-consensus and tunnel vision are some ways in which mindreading has gone astray.

The value of mirroring and mindreading is that it helps us to anticipate each other's behaviour. Through perspective taking, we can estimate what someone else is currently believing about the world, given their point of view. This knowledge can also help leaders design strategies to successfully interact with followers and motivate them.

The war for kindness

Being socially aware requires that we practice empathy, which means being able to put ourselves in another person's shoes, sense their emotions, and understand their perspectives. Empathy is in short supply and isolation and tribalism are rampant. We struggle to understand people who are not like us but find it easy to hate them.[207] Empathy is derived from the German word *Einfühlung*, which means "feeling into". Empathy is not the same as condoning others' beliefs – it merely entails trying to understand them better.

Empathy calls on many of the executive functions of the brain. It requires inhibitory control or inhibiting our own thoughts and feelings to consider the perspectives of others; cognitive flexibility to see a situation in different ways; and reflection or the ability to consider someone else's thinking alongside our own. People who have empathy can effectively focus on others and are easy to recognise. They are the ones who find common ground, whose opinions carry the most weight, and with whom other people want to work. They emerge as natural leaders, regardless of organisational or social rank.

The ability to show empathy and mentalise, as well as to try to interpret and clarify what team members are saying (the rational) and not saying (the irrational), is a key social-relatedness ability that anybody, and especially leaders, should cultivate to build a well-functioning and neurologically integrated team.

Much has been written about empathy. Empathy is regarded as an important constituent of social cognition that contributes to our ability to comprehend and respond adaptively to others' emotions, succeed in emotional communication, and promote pro-social behaviour.[208]

Despite widely varied conceptualisations and apparent incongruencies, there is acknowledgement that empathy contains both an affective component, namely, subjective experiences of and responses to the emotions of others, and a cognitive component, being the ability to understand others' motivation.[209, 210]

Do not get carried away with empathy

Another way to look at empathy is through the lens of the construct of compassion. In a study, volunteers either received empathy training (focusing on feeling the pain of someone in distress) or compassion training (focusing on a feeling of warmth and care toward that distressed person). The former group generated the typical neuroimaging profiles, including heavy amygdala activation and a negative, anxious state. Those who felt highly distressed would rather tend to their own needs, instead of the needs of others. Those with compassion training showed activation in the (cognitive) dlPFC, coupling of activation between the dlPFC and dopaminergic regions, more positive emotions, and a greater tendency toward pro-social behaviour.[211]

The punchline is that empathic states are most likely to produce compassionate acts when we manage a detached distance. Thus although an empathy deficit is not healthy, too much empathy is also not the way to go.

#Neuro-hacks to enhance understanding of self and others

- **Fine-tune your empathy**. Develop your **affective/feeling empathy**, also called emotional empathy, i.e. the capacity to respond with an apt emotion to another's mental state. The ability to empathise emotionally is supposed to be based on emotional contagion, i.e. being affected by another's emotional or arousal state.[212]

- Show **empathic concern to others**. This means showing other-orientated feelings of sympathy and concern. A behavioural marker could be, "When you notice that a colleague seems upset, instead of going on with your work, stop and ask them why".

- **Develop your cognitive empathy** or the capacity to understand another's perspective or mental state. This is also called perspective-taking and reflects the tendency to take the psychological point of view of others. A behavioural marker of perspective-taking or the tendency to take another's point of view includes: "When I am upset at someone, I usually try to 'put myself in his shoes' for a while."

- **Be aware** that too much empathy leads to **personal distress** where feelings (self-orientated anxiety, etc.) get in the way of helping or experiencing others in distress. A behavioural marker could be: "In emergency situations, I feel apprehensive and ill-at-ease."

- **Know thyself:** Oxytocin makes us suspicious of strangers, so empathy does not particularly translate into compassionate acts towards perceived strangers.

- **Work on collaborating instead of competing**. When you get stuck, ask colleagues, "What am I missing?" Listen to the answer.

- **Ask questions that get to a deeper level**. The intention here is to get to the personal story and identity behind the answer. We all relate to stories of human imperfection, rather than a blueprint of a best practice. "What was your favourite meal as a child" will get you deeper into connecting than asking, "What is your contribution to the team's morale?"

> **# Neuro-Insight: "Meanwhile, back at work... the good, the bad and the ugly"**
>
> Matthew was dreading his meeting with the Executive Director (ED), whose secretary had advised Matthew that the ED wanted to speak to him about the allocation of funds for international travel. The ED's international travel budget was depleted and he wanted Matthew to re-allocate funds that had specifically been earmarked for programme development into his travel account.
>
> Upon arriving at the meeting, the ED asked Matthew how soon this fund re-allocation could take place. When Matthew informed him that this could not take place without first obtaining approval from the donor of the earmarked funds, the demeanour of the ED changed dramatically. Frowning and visibly upset, he looked at Matthew and said, "I expected much better cooperation from you and find your response to this request completely unacceptable.
>
> "The public relations that I build during these international travels are vital to advertise the work of the organisation and to enhance its reputation in the international development aid world, and I therefore now instruct you to see to it that the requested funds are re-allocated with immediate effect." When Matthew once again reiterated that the funds under his control were earmarked for other purposes, the ED abruptly closed his diary, angrily announced that the meeting was over, and asked Matthew to leave his office, stating that he would find other ways to have the requested funds re-allocated. Matthew left the meeting feeling highly distressed that he had let the ED down and he struggled to focus on his work for the rest of the day.
>
> **Do not make bad situations worse**
>
> Whatever you think of the content of the ED's meeting with, and his instructions to, Matthew, it is clear that how the ED handled the meeting made a bad situation worse. Missing from the ED's behaviour were the use of self-awareness, self-management, and empathy that are key to social savvy. If the ED had better understood and applied mirroring and mentalising he could have handled the meeting with Matthew more skillfully.

Self-awareness and contagious feelings

When the ED became visibly upset in his meeting with Matthew, he was engaging in something called emotional contagion. Whenever we interact with others, our brains and bodies react to the feelings of those around us. These reactions involve neural circuitry like the mirror neuron system that operates in emotional contagion. Those systems work automatically, instantly and unconsciously, and are beyond our intentional control. Studies have also shown that people are hardwired to pick up signals from someone else. When it comes to spreading emotions, those who have the most social power and influence become the "sender" of feelings. The ED did not realise that he was spreading his frustration and anger, and negatively impacting on Matthew's subsequent feelings and behaviour.

NOTES

_____

_____

_____

_____

_____

_____

_____

"Over 90% of our behavior is driven by our unconscious motivations, values, assumptions, beliefs and habits."

John A. Bargh[213]

5. Sustained collaboration in groups: The interconnected self, team, organisation, culture

"We all have the same universal desire, which is freedom from suffering and pain. Our well-being depends on rising together and looking after each other."

Yeshe Dawa[214]

From only 'me' to also 'we'

The global trend towards team-based organisations is growing. Multi-functional networks that work in an agile way and that need to scale quickly to demand, require a collaborative approach. This collaboration approach has similarities to the African "Ubuntu" philosophy that captures the essence of being human; it means that a person is a person through other people. It embraces generosity, caring about others, and being willing to serve others for the sake of the greater good. As Nelson Mandela said, Ubuntu is a vital instrument to "bridge the gaps between people in the workplace, stakeholders within and outside the enterprise and business, and the broader society in which they operate" (cited in Stout-Rostron[215]).

Our life at work occurs in the context of social interactions and it will continue to do so. It requires the ability to take into account social norms and understanding others. The importance of social interactions cannot be underestimated. As John Cacioppo aptly put it: "Human survival depends in large part on the formation of alliances and accurate judgements."[216] It is well understood that as human beings, we are a social species, so much so that the survival and success of our species depend on our ability to function in complex social interactions.

Psychologists have long contended that people exhibit in-group (us) and out-group (them) biases.[217, 218] This process can be conscious or implicit. In fact, people exhibit a natural tendency to associate with those who are like themselves and maintain some type of social distance from those who are dissimilar. To some extent, this natural tendency helps satisfy certain basic psychological needs in people who cooperate with in-group members.[219]

Perceptions of in-groups and out-groups (us/them)

Us/them-ing is about relations between groups and our spontaneous tendency to favour in-groups over out-groups. Similarly, hierarchies are about a kind of relationship within groups, namely our automatic propensity to favour people close in rank to us over those who are distant.[220] These tendencies appear early in life. The intertwined cognitive and affective underpinnings are a primate brain inheritance and our brains are inherently quick to spot us vs. them. The self-regulation to take a pause before deciding if someone is a friend or a foe comes online much later. To harmonise with the group that we want to be part of requires self-control in favour of social norms, ways and values. Deploying the brain's braking system mediated by the ventrolateral PFC helps us support the group, sometimes at the expense of our own unsocialised instincts or impulses. In essence, self-control is the experience of applying effort to overcome something.[221]

Survival mechanisms underpin in-group/out-group distinctions. In our desire to feel safe, we bond together with those whom we see as most like us so that we can protect ourselves from those who might do us harm. The virtual barriers we build keep the outsiders away and allow us to go on with our daily lives, feeling protected and secure. However, it is precisely these barriers that keep us from bonding with our fellow human beings and in this way, undercut our true social connectivity and feeling secure.

This in-group positivity and out-group negativity also impacts our natural tendency for empathy. Our level of empathy can be impacted by the extent to which we perceive others to be like us (in-group), as well as our prior experience with an individual relative to levels of fairness, and therefore trust. We empathise to a greater degree with those who are similar to us and who we believe are fair and trusting.[222]

In a study by Doyle, Srivastava, Goldberg and Frank[223], it was found that a person's base rate use of the first-person singular "I" or plural "We" might indicate their degree of group identity internalisation. This study showed that people who do not switch from "I" to "We" pronouns during the first six months of employment are more likely to leave their jobs. Not feeling part of the in-group is regarded as being one of the strongest predictors of turnover.

From foe or frenemy to friend

Given that thinking about self and others is inevitable, one way to help reduce the time needed for a new or changing team to be fully productive, while minimising the tension, fear, or anxiety common in group development, is for the leader to ensure that the team engages in ways that will bring to the surface points of resemblance, strengthen resonances, cultivate empathy, and contribute to feelings of trust. Only after a sense of relatedness or seeing others as part of the 'in' group has been established, can social differences be effectively addressed.[224]

Social support has neural underpinnings – in healthy adults, perception of social support is correlated with increasing MPFC thickness and is negatively correlated with amygdala volume, brain regions that are associated with resonant relationships.[225]

In neurofeedback training, an individual can be entrained to get into "being in the zone". When two or more people are socially engaging with one another, that too appears to involve entrainment – or being in the same zone. This shared entrainment shows up as neural synchrony on EEG measuring. 'Being in the zone' or having 'brain-to-brain synchrony' is a simple function of our mutual attention to a common social dynamic stimulus. Feeling that you belong, actively reaching out to include all others and shared goals enhance this neural synchrony.

Gut feelings and mental short cuts

In an uncertain world, we sometimes have to ignore information and rely on our brain's short cuts, heuristics, rule of thumb, social norms, implicit and explicit biases, and gut feelings to make decisions about our world at work.[226]

One may hold negative and positive attitudes toward others without overly expressing them. An expressed attitude is one that people report.[227]

Cognitive biases are systematic errors in thinking that affect the decisions, judgements and conclusions that individuals make.[228] Sometimes these biases are related to memory.[229] Thus, the way you remember an event may be biased for several reasons and that, in turn, can lead to biased thinking and decision-making. In other words, a cognitive bias is a mistake in reasoning, evaluating, remembering, or some other cognitive process, often occurring because of holding onto one's preferences and beliefs regardless of contrary information.

Social cognition theorists suggest that cognitive bias is a tendency to process information by filtering it through one's personal likes, dislikes and past experiences. Contingent with our personal perception of the specific input, we create our own unique subjective social reality and it is this "construction of social reality" and, not the actual objective input, that therefore may determine how individuals subsequently behave in the social world.[230]

However, it is argued that cognitive biases or heuristics can also be seen as rational in an underlying sense in that they are rapid, can be made without full information, and can be as accurate as more complicated procedures.[231]

Why do gut feelings and ultimately bias exists?

Our implicit bias grows from our early experiences and models of behaviour. As children, we observe the adults around us to inform our own values, beliefs and perceptions. In addition to forming implicit biases, humans instinctively sort into "in-groups". This happens when people preferentially identify with a group, based on factors such as race or ethnicity. Our implicit biases further consolidate affiliation with our in-groups. On the surface, these associations can create feelings of security and comfort.

Merriam-Webster's Online Dictionary[232] defines bias as:

- "a BENT or TENDENCY
- an inclination of temperament or outlook; especially a personal and sometimes unreasoned judgement: PREJUDICE".

In turn, the American Psychological Association Dictionary[233] defines "in-group" bias as "the tendency to favour one's own group, its members, its characteristics, and its products, particularly in reference to other groups". Thus our bias is influenced by our unconscious drivers of how we interpret the world. It is evolutionary, adaptive and prevalent in many of our decisions, so it is fair to say that we are going to have a biased mental model of seeing the world.

In essence, bias implies that we are making thinking errors, so why do we overlook our mistakes? Because being wrong is painful and being right is rewarding (there are many dopamine pathways in the brain – see Figure 26), and we feel right far more often than we should.

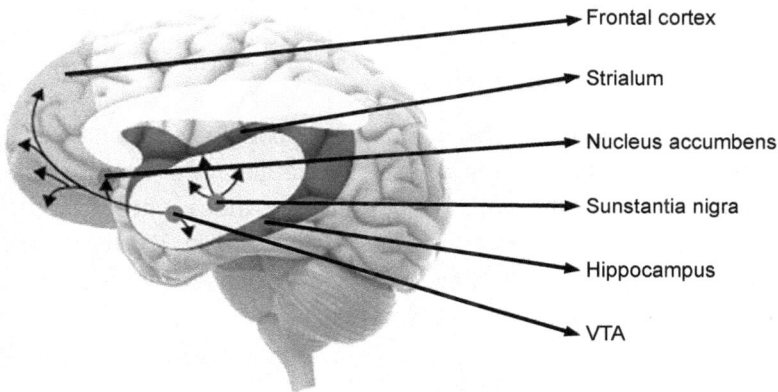

Frontal cortex

Strialum

Nucleus accumbens

Sunstantia nigra

Hippocampus

VTA

Figure 26: Dopamine pathways in the brain (Source: Telzer)[234]

The neural basis for gut feelings and bias

The dual system of reflexive and reflective thinking modes discussed earlier could help explain our understanding of unconscious bias from a neuroscience perspective. An implicit attitude (related to the reflexive system) is rapid, automatic and comprises unconscious evaluations in response to stimuli, whereas an explicit attitude (related to the reflective system) is a relatively slower, deliberative and conscious evaluation, based on contextual information.

Mirror neurons might also help explain discriminatory behaviour. To some extent, they could represent the deep causes of discriminatory behaviour because they reside in the "deep-brain structure of the limbic system".[235]

"To a closed-minded person, an open-minded person looks biased.
An open-minded person may think the views of a closed-minded person are biased."

Unknown

There are many biases (see Appendix D for a review of how to spot and tame them), but for now I focus on two unconscious biases that every one of us has.

The insider bias: "Beware of strangers"

Our brains are wired to recognise and respond to threats to our survival, and to immediately recognise subtle cues regarding whether another person is part of what we regard as our "in-group" or a potentially hazardous "out-group". This is why we cannot just all get along as every one of us has an **unconscious bias** – we prefer individuals who look, act and sound like us, and we tend to care more about people who are part of our "tribe".

This mental short cut is observable in infants as young as six months old. No matter how intentionally we claim to be inclusive, accepting and open to all, our minds are habituated to label others so instinctively that those who do not look like us are recorded as a threat. This conclusion happens within 200 milliseconds. These embodied unconscious biases, while valuable to our ancestral survival, limit our ability to empathise with dissimilar others.

The science of unconscious bias

Researchers explored mentalising (thinking about self and others) based on whether the other individual is similar or dissimilar. They found that mentalising about a similar other (us) engages a region of the ventral medial prefrontal cortex linked to self-regulation, whereas mentalising about a dissimilar other (them) engages a more dorsal sub-region of the medial prefrontal cortex. This implies that we use the self as reference when thinking about similar others[236], and because we tend to view ourselves favourably, we tend to be more effective at recognising other faces when they are similar to us and specifically the same race as us.

This has real-world consequences and raises the question – how do we look beyond our stereotypes and biases?

Time for inclusion and diversity: Life's better together

Diversity is not only about differences; rather, diversity refers to both the differences that help us see each other as unique individuals and the similarities that help to connect us. These differences and similarities can be things such as age, race, gender, sexual orientation, ethnic background and physical ability. But diversity also covers other aspects of our identity – gender identity and expression, core values, cultural norms and the ways we approach others and process information.

The goal of growing diversity in our workplaces has been around for a short time. The South African Bill of Rights, which is part of the Constitution of 1993, has contributed to increased awareness and acceptance of differences, equality and freedom of association. However, to move from laws and rulings about diversity and inclusion to the real asset and business case for a diverse and inclusive workforce, we need much more social connectivity.

Simply putting disparate people together might tick the diversity box, but it is simply not enough. That is why the idea of inclusion, or creating an atmosphere that values, respects and intentionally engages differences, is so essential. An inclusive culture is one in which all employees feel comfortable, connected and supported with individuals who are similar and those who are different. They are free to express their opinions and disagree because there is a high level of trust among all group members.

Studies have shown that constructing diverse teams in inclusive settings can noticeably improve problem-solving and increase employee innovation and engagement. Inclusion is the key to unlocking and maximising the potential of the workforce.

According to Nene Molefi, a thought leader in the diversity and inclusion space, inclusion is a call to action[237] at three levels:

The first level of the call to action speaks to leaders, who should be modelling the way of inclusive behaviour and deliberately looking for exclusion in the workplace.

The second call to action speaks to those who are excluded; you must speak up and include yourself. Inclusion starts with the word "I". It requires being constantly vigilant, and actively looking out and reaching out to those who are being excluded and then actively including them. This also means including yourself by actively speaking up when you feel excluded and then engaging with others about it. Your boss does not lie awake wondering if you are included or not; if you feel excluded, you must recognise that others on your team may be oblivious to what you are experiencing.

The third call to action is to the observers: those who are not being excluded or doing the excluding, but through the power of proximity and position they can do something about it.[238] These things do not just happen automatically – we have to make them happen.

129

Inclusion is a call to action for all to become activists for what is right – to simply affirm the desire to work in a culture of fairness, justice and inclusion.

Dialling down bias and dialling up inclusivity

- Learn to interpret and articulate your own biases to enable a shift from often fragile work dynamics to those of flexibility and flourishing.

- Develop your intrinsically inclusive mindset. Become curious about others and wanting to learn more about them, especially when it comes to interacting with those who are different.

- Widen your "in-group" circle by creating new markers of membership such as clothing with the company logo, team pictures and invitations to corporate events.

- Create shared experiences and goals. Shared purpose is a foundational social driver that occurs because we are wired to be part of something greater than ourselves – provided we create and believe in that "greater than the self" cause.

- Compensate for implicit preferences – **consider what gets into your mind in the first place**. For example, this could mean going out of your way to watch television programmes and movies that portray women and minority group members in positive or counter-stereotypical ways.

#Micro-behaviours are tiny, often unconscious, signals, jargon, facial expressions, tone of voice and even postures which can influence how included (or excluded) the people around us feel. The term 'micro-behaviours' was created by psychologist Mary Rowe from MIT. These minute actions can result in immense pain and problems. Since these behaviours are 'micro', they are mostly regarded as insignificant, yet micro-behaviours can significantly impact the level of inclusion and value we feel. It is a form of subtle discrimination and is one of the most significant scaffoldings for segregation in small groups to an entire society.[239]

Also watch *Be colour brave not colour blind* – Melody Hobson's TED talk.[240]

Negative micro-behaviours can lead to mega problems resulting in cumulative patterns of pride, prejudice and pain (a downward spiral of languishing). Positive micro-behaviours have a ripple effect and lead to an upward spiral of flourishing.

 # Neuro-Insight: Spot your micro-aggressions and micro-affirmations

Reflective questions to gauge your markers of micro-behaviours

Negative micro-behaviours (or micro aggressions):

- Are you flippant towards people more junior to you, e.g. "I didn't have time to read that email"?
- Do you use nicknames for some people and not for others?
- Do you pronounce people's names incorrectly?

Positive micro-behaviours (or micro-affirmations) include:

- Do you pay attention to someone when they are talking?
- Do you let someone finish their sentence?
- Do you make eye contact?

NOTES

_____

_____

_____

_____

The false-consensus bias

Attention is a finite resource – this means that a high cognitive load, like solving business problems, reduces a leader's ability to be socially aware or to be empathetic towards others. Given that leaders are often under tough deadlines to churn out profits (and are thus under high cognitive load), they default to their own 'know-how' or mental short cuts, and even fall into the false-consensus bias. This can have negative consequences for both business results and team effectiveness.

To regulate the exhaustion of compounded, challenging decisions, the brain resorts to heuristics and cognitive rules-of-thumb, which are developed over time. These mental short cuts are convenient but can also distort a leader's clear thinking. While it is difficult for leaders to detect and take control of their cognitive thinking errors, they can apply rational thought to detect others' faulty intuition, improve their decision-making, and reduce tunnel vision.

When we are under severe stress, the surges in our neurotransmitters and hormones, specifically nor-adrenaline and cortisone, affect our reasoning and cognition. Since stress is contagious, we and those around us will react to workplace stressors by falling back on the status quo of doing things, as under elevated stress we do not have the mental bandwidth to reason in a novel way.

As mentioned before, our social cognition system is inversely related to our analytical system. Thus a high cognitive load (analytical task-driven roles) reduces a leader's mentalising ability (understanding the minds and intentions of others, perspective taking, and shifting attention to focus on the needs and values of others). It requires deliberate focus and effort to peer into others' minds while solving analytical problems.[241]

This enforces why it is important for leaders (or anyone) to keep their minds clear of clutter.

#Neuro-hacks for growing the interconnected self, team and organisation

- Be aware of **the experience you create for others**. Formal organisations have social structures and social experiential factors where leaders (disproportionately) influence the behaviours of others by either creating pleasurable or painful experiences.

- Build **micro-positive actions** (small gestures that act as social signals), for example when a colleague joins a conversation. Take a moment to bring them up to speed. Avoid **micro-aggressions** (indirect and unintentional moments of exclusions).

- Recognise the **arbitrary nature of many in-group/out-group distinctions**. The example of pedestrians and motorists serves to demonstrate this point. Your in-group at one moment is your out-group the next.

- Lead by **inclusive example**. Full stop.

- Take **a first-person account** of others' perspectives, ideally in a written form. Thus, imagine for a moment that you are this person, walking through the world in their shoes and seeing the world through their eyes. Think about how you, as this person, would experience this event.[242]

- Maximise **face-to-face interactions**. Make a regular effort to engage in social interactions in the workplace, such as regular lunches. It builds in-group dynamics. Try to be around people who see things positively.

- **Compensate for your implicit preferences**. For example, if you have an inherent preference for young people, you can try to be friendlier toward elderly people.

- Look for **commonalities** (not differences) between opposing groups. Fans of opposing sports teams equally love the sport. People of different religions regard their faith as important to them. There are basic human needs that transcend particular labels. *#Be colour brave not colour blind.*

- Work on constructing your **inner sense of security**. We are more likely to stereotype when we feel threatened. If you feel more assured about your own identity, chances are slim that you will criticise someone else's.

- Try to **surface assumptions** that impact your decisions to discover some of your own biases, or take a short test at www.implicit.harvard.edu.

- **Investigate the 'why'** of a decision and what criteria you used to get there.

- Use **generative language**. If someone has an interesting suggestion, respond with, "Let's try it". If you like the essence of someone's idea, say, "Building on that idea...", or make, "Yes and..." your catchphrase.

- Grow awareness of your **emotional bias**, i.e. the degree to which your nonconscious negative biases impact your thinking. Mindfulness training can dampen your negativity bias.

> *"Odd as it may seem – I am my remembering self, and the experiencing self, who does my living, is like a stranger to me."*
>
> Daniel Kahneman[243]

Conclusion

Our social capacity is the capacity for building connections and keeping relationships. If this is high, in safe environments we allow others to see the "heart" of our decisions. Social comparison, social isolation and chronic loneliness can kill us. The brain is selective and takes many short cuts, therefore we need to make social connections, inclusion and diversity our number one priorities.

6.7 Stratification of purpose and goals – the will, the way and the habit

"Inspiration is not a random event but a reproducible state that happens at the cusp of support and challenge."

Dr John F. Demartini

At a Glance

We are wired for purpose, meaning and goal pursuit. In a rapidly changing world, increasing complexity and uncertainty can undermine our basic human need for control, which can result in disengagement and even a downward behavioural spiral. This is where meaning, purpose and goals can provide a stress buffer and an elixir for change and flourishing at work.[244] A variety of factors play a vital role in flourishing in the world of work like role fit, trusting relationships with managers and colleagues, as well as clearly defined goals and role clarity.[245]

The neuroscience of purpose and goals is a big and emerging field. Discussing Goal Theory (cognitive representations of future outcomes) and Motivation Theory (processes that energise and direct behaviour) are beyond our focus here.[246] We focus here on how neuroscience can contribute to understanding goal pursuit at work. We take a broader field of view and also include related constructs like intentional change, meaning, purpose and motivation. We focus on:

1. **Change in the Brain** – Most goals are set to achieve a change from the way things are;

2. **Goal-Directedness** – The unconscious brain state necessary for survival;

3. **Goal Pursuit: The Will** – Our motivation to pursue goals driven by a will for personal and social meaning and purpose; and

4. **Goal Pursuit: The Way** – The means to pursue goals or doing what it takes to achieve the goal.

| 1. Discontent with the status quo "From DUH to AHA" | 2. Goal directedness "Seek and you shall find" | 3. Motivational processes "The will" | 4. Cognitive processes "The way" |

1. Discontent with the status quo: From DUH to AHA

"Imagination is more important than knowledge. For knowledge is limited, whereas imagination embraces the entire world, stimulating progress, giving birth to evolution"

Albert Einstein[247]

The quote from Einstein might sound like a sweeping statement about innovation being better than knowledge, but it also talks to how we integrate knowledge to evolve and expand; to move forward – a core ingredient of flourishing. The human brain has the unique ability to take what it knows and dream up something new. There are two main areas where neurobiology affects innovation. First, our brains are built to resist change. In order to conserve the status quo, the brain generates feelings of discomfort when we try new things or attempt to change. The "fear factor" is a root cause of organisational and personal failure. This is counterbalanced by other systems, driven by dopamine and opiates that reward exploration, discovery and the achievement of something new. To risk it into the unknown, we need to be at the cusp of support and challenge. It's a balancing act – when we experience too much stress and threat, the tendency is to withdraw into habitual known responses. When we feel sufficiently (but not overly) secure, we venture into new territory. The creative process and innovation are often born out of discontent with the status quo. Distress and disappointment are a breeding ground for insight, provided we slow down our thinking.[248]

Insight follows a pattern: explore a topic, gather the data, work on a solution, take a break, have a breakthrough, and work to make the insight into something functional. The actual process of creating new insights (a precursor to innovation) has been shown in electroencephalography (EEG) technology that measures the electrical activity on the brain's surface and shows five levels of activity, measured in hertz (Hz). Two seconds prior to insight, there is a spike in the alpha wave, 10 Hz, which at the moment of insight, flip flops with a gamma band spike (40 Hz).[249] The 40 Hz frequency is known as the binding frequency, i.e. the rate at which neurons that fire together, wire together. It seems that the moment of insight signals the coming to consciousness of a newly integrated cognitive map; a momentary binding of a novel neuronal pattern – creative thought. Research has shown that the gamma band wave spike, and subsequent adrenaline/dopamine rush at the moment of insight, is fleeting, perhaps only 10 minutes. Consequently, it can be said that insight is short term and provides motivation to act.

By clarifying the brain basis of insight, we can cultivate creativity and innovation by acting as thinking partners, by valuing unstructured time as a key factor in innovation, and by being deliberate in asking reflective questions that facilitate the neural connections in others' minds.

Make time and space in your diary (and your team's) for sensitive reflection, and know when to switch from analytical (prefrontal cortex) to quiet mode (take a walk, listen to music, doodle) in order to move beyond mental impasse to insight.

Inducing optimism also broadens creative thinking and highest performance.[250] The root word of optimism is from Latin *optimist*, meaning the best – the superlative of good. Essentially, it's the tendency to have confidence in the future.

When you need to come up with an ingenious solution, sleep on it. The power of slumber should allow you to return to a complex problem with renewed mental vigour and improve your chances of achieving a solution.

However, having AHAs are not enough to change behaviour – neuroscience reveals the degree to which implicit memory created by biological responses to sensory, social and emotional cues dictates behaviours. Thus, learning new information alone typically can't change our impulses. We need to have new experiences that have enough "pull" (read purpose) and we need repetition (goals and actions that are congruent with purpose). So, in essence, we generate new ideas when we combine an open mind with expertise. Innovators are persistent and push past defeats or disappointments. They show Purpose and Grit.

Another key driver in innovation is diversity in thinking. Diverse companies are more likely to be innovative, especially when they combine inherent diversity with the diversity that they acquire through varied experiences.

A diverse workforce adds a variety of perspectives. Broadening employees' perspectives nudges them to seek wider information. Having teams from different disciplines interact promotes innovation.

> *"Culture, Engineering, Art, Science, Music and Technology: These things are only possible because we can make things up."*
>
> Unknown

Incongruence – the cornerstone of change at work

Incongruence forms the heart of life, survival and flourishing; it is the cornerstone of change. At its core, effective goal pursuit is about change. The brain works in a protective way, typically resisting change. Therefore, any goals that require significant behavioural change or thinking-pattern change will automatically be resisted. Controllable incongruence is facilitated when we are confronted with triggers (like new goals) that are manageable and adaptable. In controllable congruence, the cognitive brain is recruited and executive functions are activated, resulting in effective self-reflection and problem-solving capabilities. Uncontrollable incongruence is facilitated when we are confronted with challenges like goals that are perceived to be unachievable – it activates the stress response with diminished activation of the frontal neural systems.[251]

This is similar to the concept of creative tension vs. emotional tension – essentially a structure that helps to facilitate creativity and change.[252] You create creative tension when you clearly and succinctly articulate your vision and your current reality and the gap between your vision and your current reality becomes apparent. Our brains do not like the cognitive dissonance in saying "this is what I want" and recognising that you do not have it. It is highly motivated to relieve that cognitive dissonance by closing the gap between current reality and vision, and so this allows one to release more emotional and creative cognitive energy into finding ways to close that gap.

The Salient Network (with Anterior Cingulate Cortex [ACC] as key node) is involved in detecting discrepancies between the desired state and the actual current state of the subject matter. The salience network also decides which things you pay attention to and which things you ignore.[253] When the error detection circuitry of the ACC fires too often, it brings on a state of anxiety or emotional tension where the sub-cortical emotional system then takes over, resulting in feelings of becoming overwhelmed or wanting to discard the vision (uncontrollable incongruence). As mentioned before, there is an optimal or flow state where the arousal levels (catecholamine – dopamine and norepinephrine release) are "just right" to release optimal prefrontal cortex abilities like focused thinking and vision achievement (controllable incongruence).

2. Goal-directedness: Seek, and you shall find

Goals are at the heart of organisational change and performance. The objective of having goals is to provide an endpoint or target, against which we can determine performance and/or learning and development.– successful or otherwise. Setting goals is a clear prerequisite for measuring their attainment. We aspire to goal achievement – goals represent our progress over time and when achieved, provide a sense of completion and satisfaction. In this way, goals are integral to giving meaning and purpose to our lives. How would you measure performance or learning at all, if there were no goals?

To achieve individual, team and organisational purpose and goals, we need to understand goal-directedness first. Note that goal-directedness is not the same thing as setting goals. As goal-directedness is an unconscious brain state essential for goal achievement, it is of special interest at all levels in the organisation. Goal-directedness, developed for survival, initially centred on basic physiological needs like food and shelter.

Goal-directedness in its basic form is a subconscious process in the brain that is associated with the reward system. This was described by the founder of the field of affective neuroscience, Jaak Panksepp[254], as the 'curiosity' or 'seeking' system. Goal-directed behaviours involve risks, costs and rewards. An evolutionary example: straying from the herd may offer zebra better opportunities for grazing but, at the risk of becoming an easier target for a prowling lion. Attacking the zebra offers the lion the promise of a meal, with the risk that his energy resources will be depleted if the zebra gets away. Thus, the neural mechanisms responsible for goal selection must weigh up anticipated risks, costs and rewards of behaviours that are likely to attain a specific goal. The brain's reward circuitry may provide a common logic for goal selection – dopamine and opioids get secreted when we achieve something we really want.

With the evolution of a much more sophisticated and creative brain, we have developed the capacity to consider (using conscious cognition) the imagined options that shape goal-directed behaviour for survival in a more complex world. Thus goal-directed behaviour enables a flexible, dynamic and rapid adaptation to our internal motivational states and environmental conditions.[255] Examples are: building a world-class team, revolutionising an industry, or raising happy children.

An unhealthy state of goal-directedness talks to immediate gratification and dopamine cravings. Why be motivated to achieve a long-term goal when you can get that same satisfaction of fulfillment from social media or other immediate gratification habits (often without leaving your seat)? Dopamine causes you to want, desire, seek out and search. It intensifies your general level of arousal and your goal-directed behaviour. However, it is the opioid system (separate from dopamine) that makes us feel pleasure. This requires an understanding of liking vs. wanting.

The neurochemistry of goal-directedness: Liking is for Wanting[256]

Wanting and liking something are separate urges controlled by different brain circuits. 'Wanting' is driven by large pathways in the brain that use the neurotransmitter dopamine, while 'liking' is controlled in smaller, pleasure-generating centres that do not use dopamine. These two pathways are complementary. (The wanting system propels you to action and the liking system makes you feel content and therefore pauses your seeking.)[257]

When the brain realises that an action might lead to a reward, dopamine is released to motivate you to act. Dopamine only gets released before an action is taken, not during or after. If your seeking isn't switched off at least for a little while, then you start to run in an endless loop. The dopamine system is stronger than the opioid system. You tend to seek more than you are satisfied, and seeking is more likely to keep you alive than sitting around in a fulfilled lethargy.

On a practical level, the central problem that goal-directed behaviour focuses on is making why, what, where, when and how choices (H4W). This is a vast and complex field of study which is beyond the scope of this book.[258] To simplify this I focus on the 'will and the way' of goal pursuit, referencing some of the brain networks and concepts involved..

Core brain networks that are linked to the 'will and the way' of goal pursuit

The human brain is massively interconnected. To explore motivation and goal pursuit further, I refer to three brain networks: (1) The reward network (our everyday behaviour is shaped by reward, which is both tangible and intangible external stimuli that lead to satisfaction); (2) the valuation network (the personal quantification of the benefits or costs associated with any action); and (3) the self-regulation network (controlling impulses, or temptations that compete with a goal).[259] I will refer to these as we go along.

3.   Motivational processes involved in goal pursuit – the 'will'

the motivational processes of goal pursuit are about the 'will' that propel us to change behaviour.

This involves connecting goals to a sense of personal meaning and purpose, the ideal self-concept, the greater good and pro-social behaviour. Motivation has real value in the world of work; it produces results.

Wired for purpose and the greater good

Victor Frankl was the first to identify purpose and meaning as motivators to change, or as the drivers used to overcome circumstances.[260]

> *"Man's main concern is not to gain pleasure or to avoid pain but rather to see a meaning in his life. That is why man is even ready to suffer, on the condition, to be sure, that his suffering has meaning."*
>
> Viktor Frankl[261]

In essence, purpose is our reason for being or doing what we're doing; our usefulness. Certain people appear to be motivated and engaged in achieving an aspirational purpose or personal goal such that it aligns their efforts, their thinking, and their decision making. Others are motivated by more immediate passions. Purpose is described as a centralizsed, self-organising life aim that organises and stimulates goals, manages behaviours, and provides a sense of meaning.[262] "Meaning and purpose are separate, although highly related, constructs that build off of one another so as to contribute to the broader concept of the 'good,' or meaningful life".[263]

In deconstructing these two concepts, it is found that meaning is linked to an integration of the past, present and future, whereas purpose is a future-directed part of meaning that may not integrate past and present.[264] However, if past and present are integrated into purpose, our lives fall into place. We know that everything and everybody helps us grow towards achieving our purpose. In this way, purpose and meaning both drive our actions and express or define our identity.[265]

Martin Luther King Jr. had an inspiring definition of purpose: "Leading from purpose means making a purpose more powerful than yourself. Leading from purpose means valuing contribution over personal achievement."

Purpose, not happiness, provides enduring fulfillment.[266] So, purpose is the cause and happiness the effect/result. Purpose, when rooted in something outside ourselves, constantly pulls us forward and through difficult times. When we're pulled by purpose, we're more likely to reflect on our past and future and are much more likely to regularly consider our impact on others.

In addition to being future-directed, the construct of purpose is becoming more understood to include doing something that a person feels driven to do in which the benefactor or benefactors are not themselves.[267] Purpose is also characterised by a cause that is personally meaningful and helpful to the greater society. Altruism has allowed us to survive as a species by compelling us to help one another. So while we're wired to set goals, we're also wired to be useful to others. The part of the brain that is associated with altruism is primitive and has allowed us to survive as a species by hard-wiring us to get on and support each other. It's not a surprise, then, that we experience significant neurological benefits when we set goals that are useful to others. This is also called the "helper's high", which consists of positive emotions following selfless service to others.[268] However, the catch is that the goal needs to be specific or concrete to deliver this happiness to the giver. So having a goal of "supporting environmentalism" (an abstract goal) does not provide the same helper's high as a more concrete goal of "increasing the number of plastic resources that are recycled or reused".

The specific brain areas involved in prosocial behaviours are mostly located in the insula[269], which may encode intangible benefits of prosocial behaviours beyond pain-pleasure calculations. In this way we can replace the word 'purpose' with 'usefulness' and 'helpfulness'. Setting concrete purpose-driven goals and working towards your purpose every day produces a rush of oxytocin, dopamine and serotonin – what neuroscientists call the "happiness trifecta" – causing a boost in mood and a reduction in stress and anxiety. And the happier we feel, the more motivation we have.

To thy own self be true

Goal pursuit and achievement is accelerated when it is aligned with higher order motivations, our ideal self-identity, key commitments or highest values. Alternatively, as Frankl[270] put it: If the 'why' is big enough, the 'how' will take care of itself. If the 'why' is framed as a social cause, the brain's social reward system is activated. Having a cause bigger than the self is highly rewarding for the brain. It also enhances organisational commitment and engagement when organisations get employees involved in prosocial causes. When we can see how our work has an impact on society as a whole, we experience what is called 'self-transcendence'. Maslow[271] described the significance of transcendence as "the very highest and most inclusive or holistic levels of human consciousness, behaving and relating, as ends rather than means, to oneself, to significant others, to human beings in general, to other species, to nature, and to the cosmos".

According to Maslow, self-transcendence brings us "peak experiences" in which we transcend our own personal concerns and see from a higher perspective. These memorable, magic moments or peak experiences often bring strong positive emotions like joy, peace, and a sense of profound significance of "being" in the world. And, yes, we can create these memorable moment experiences. Although peak experiences often tend to present with a burst of spontaneity, we can induce memorable moments through slowing down our thinking, paying open attention, making an effort to truly connect with others or nature, and working toward and achieving our inwardly inspired goals. A very practical example of a self-transcending peak experience is having a crucial conversation like mentoring or coaching, and feeling a shift (for both mentor and mentee or coach and coachee) from feeling successful about personal achievements to feeling a sense of greater significance through adding value to each other through shared meaning making.[272]

> "The things you do for yourself are gone when you are gone, but the things you do for others remain as your legacy."
>
> Kalu Ndukwe Kalu[273]

The neuroscience literature shows that displaying prosocial behaviour is extremely rewarding to the brain. Giving is indeed its own reward. Current changes in our organisational cultures due to the transformational impact of technology are seeing the conventional command-and-control hierarchical organisation being replaced by a holacracy-powered organisation, where the focus is on aligned purpose and engagement via information networks at the organisational, team and individual levels.[274] The need for purpose and meaning is the paradox in this high-tech world; we need high tech and high touch.

The development of self-determination theory[275] contributed to our understanding of motivation in a social context; it entails that experiencing motivation and engagement by embracing someone else's vision may very well work for a while. Boyatzis and Akrivou[276] advanced this understanding by noticing that you can be perfectly content working toward someone else's goal or objectives, until you realise that your personal dreams are being compromised because this "ought self" does not match your ideal self. This wake-up call leads to feelings of betrayal, feeling disempowered, and frustration for having wasted energy pursuing the dreams and expectations of others. This produces what is called negative emotional attractors (NEA), which have an adverse effect on motivation, discretionary efforts and engagement.[277]

But what is the secret to creating genuine, lasting change in our lives?

For desired change to be anything but transitory, we need to see that change as a positive extension of our ideal selves.[278] The challenge is to move from a problem-focused approach to change to a vision-based approach that is designed to bring out the best in individuals, teams and organisations. This produces positive emotional attractors that enable a positive state of mind and goal commitment, resulting in flourishing (neural thriving).

Creating shared goals

In our world of increasing complexity there is a move towards flatter, self-managed and networked organisational models, which go with many shiny buzzwords like, holacracy, agile, pods etc. The way we interact across this dynamic, networked organisational landscape is changing fast.[279] Three core characteristics of this self-managed world of work are: (1) Teams are the structure; it's about embracing the social brain; (2) Teams design and direct themselves (they outline responsibilities, activities and overall goals, and contain highly detailed metrics for evaluating performance); and (3) Leadership is contextual and aligned with organisational vision and purpose (leadership is distributed among roles, not individuals, and people usually hold multiple roles on various teams).

The brain science argues that team members can be aligned behind the company vision and purpose through leveraging the social wiredness of the brain. This is done by creating 'in-group' behaviour through shared goals, the mixing of skill sets and a common team identity. If we see someone as similar to ourselves through shared goal creation, we are more motivated to work together towards goal achievement. In this way, minimal group influence is required by management to achieve goals.

Driving shared goal pursuit requires mentalising ability, i.e. being able to understand others' viewpoints, and promoting debate and input from others. This is more than a participative approach. It is somewhat one dimensional to think that giving employees many opportunities to voice their views (participative) about goals will give them commitment to pursue their goals. Recent organisational behaviour research proposes that a participative approach to goal setting is not sufficient to ensure goal commitment. Instead, goal commitment can be bolstered by a deliberative approach, where managers engage employees in mutual reasoning (including both competence and warmth) when setting goals.[280]

But, be aware when your goal sharing is tipping over into social loafing (not to be confused with 'outsourcing of effort').[281] The outsourcing of effort is an effective strategy in teamwork to allocate tasks to team members based on their strengths. It is an unconscious reliance on someone else to move the team goals forward and it has a positive impact on organisational performance. Social loafing is a behaviour that undermines goal pursuit, for example in the office cafeteria where some employees lounge about while others are eager to take an order. Social loafing is a deliberate relaxation of own effort. Performance scorecards are one way to curb social loafing, but being motivated from within might be more powerful.

No one moves a muscle without a motivation

Evolution has wired us for two core neurobiological motivation systems that manifest in approach or avoidance behaviour (see Figure 27).

Co-creating and co-articulating goals with team members ensures that approach (left PFC) and avoidance (right PFC) biases anchored in personalities (extraversion or neuroticism) can be accommodated, thereby enhancing individual motivation to achieve the goal. In goal setting, we can take any situation and frame it in an approach or an avoidance way.

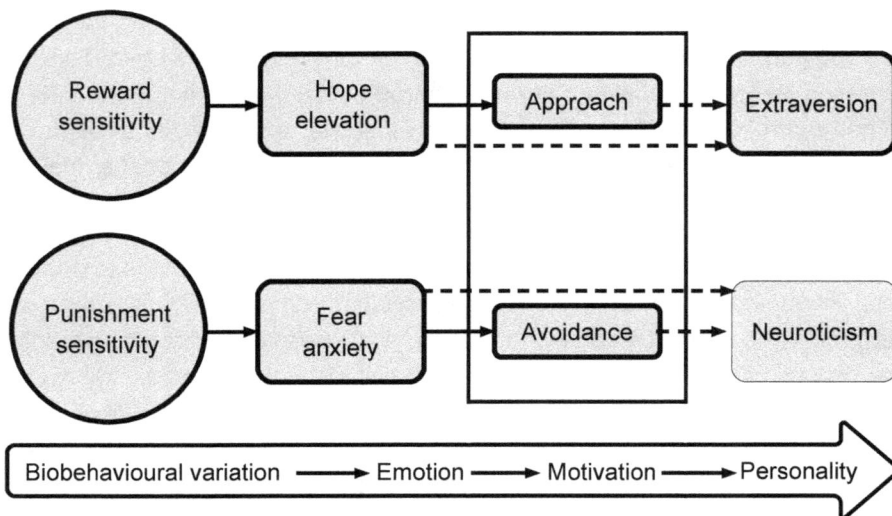

Figure 27: Approach and avoidance motivation (Source: Gray)[282]

Framing your goals with the style that motivates you (approach or avoidance) will develop your capacity to plan and organise behaviour to meet a goal. For example:

- Approach goal: To have an enriching discussion with my boss; or

- Avoidance goal: Not to argue with my boss.

We tend to have a preference for one of the two motivational systems. Those who are more approach motivated will respond more positively to an approach motivated statement, while those who are more avoidance motivated will be activated by a more avoidance type goal statement.

4. Cognitive Processes involved in goal pursuit: the 'way'

"A goal is any desired outcome that wouldn't otherwise happen without some kind of intervention. A goal is a detour from the path of least resistance."

Elliot T. Berkman[283]

Certainty (being able to predict the future) and control are primary rewards for the brain. Goal pursuit is a strategy to use to reduce certainty and control threats by giving clarity about the way forward. In fact, most accomplishments, great or small, have started with an intention that became a purpose and finally a goal.

The cognitive processes involved in goal pursuit are linked to what is known as the brain's executive functions. It is about having the skills and know-how to engage in new challenges and associated brain processes like conscious attention, working memory, impulse control, planning and habit formation.

A neural hierarchy of goals: the 'why' and the 'how'

At an organisational level, aligning individual purpose with corporate purpose is the most powerful influencer to our performance; much more so than prestigious perks. Leaders can let team members choose the 'how' of achieving goals – as this gives employees 'choice' or authority. This is a key 'towards' strategy that drives motivation and employee engagement in their work. Thus, use higher-level goals for motivation and lower-level goals for progress tracking, and make the connection between the different hierarchies of goals through goal alignment. This is done through cascading goals down from vision level to action level through company scorecard systems. In this way, a scaffolding of goal alignment can be built.

Goals can be depicted in a neural hierarchy, stretching from the motivational or abstract (the 'why' of a goal) to the 'how' or actionable part of the goal. This is called action identification theory, and these different levels of goals engage different brain systems. The 'why' of a goal engages the mentalising system in the brain. The 'how' of achieving a goal engages the motor cortex system, involving more concrete thinking. In other words, you have to be inspired and then perspire to get what you want. These should be separate components, as the brain struggles to focus on them all at the same time. Encoding personal, team or company goals on these two levels ensures stronger goal encoding and retrieval.

It can be useful to move up and down within the goal hierarchy from 'why' to 'how' in order to get unstuck from a task, or to become more practical by focusing on the 'how'. The 'how' of goal attainment involves focused attention, working memory, impulse control or self-governance, and planning.[284] This is enabled by the brain's Task Positive Network (TPN). These tasks take mental bandwidth from the brain's resources, like calculating the profit/loss on an income statement. If you love doing accounting you don't feel like it is draining your energy, but it is still a conscious, attentive process. The best way to leverage the mental bandwidth of focused attention is to dedicate specific time to it.

Develop your Episodic Future Thinking (EFT)

Neuroscience offers interesting insights into making goals come alive:

Vision is the glue that holds an enterprise together. Make it meaningful and inspiring. To ensure goals are kept top of mind, Episodic Future Thinking (EFT) should be deployed. This means imagining doing something specific in the future in a vivid, detailed way. Engaging EFT with high imagery can significantly change one's ability to focus on and achieve future goals, as it strengthens connections between the anterior cingulate, amygdala and the hippocampus.[285] In essence, this means setting goals that are aligned to your highest values and purpose and holding the image through time. The only way to hold the image (the 'why') through time is to keep it in motion (the 'how') and the only way to keep it in motion is to focus on the ever-finer details.

Pursuing your goals and changing behaviour is not easy, but is highly rewarding if it is aligned with organisational, team and individual purpose. This equates to being inspired from within (intrinsic motivation) as opposed to having to be motivated from without (extrinsic motivation).[286] When we are inspired from within, we endure both pain and pleasure to achieve our goals.

Develop your 'grit': Make choices before they become emotional

Why do some people achieve more than others? Searching for the secret to success is a major objective in the field of psychology. Over the past decade, 'grit' has been found to play an extremely crucial role in achievement at work. Grit in psychology is a positive, non-cognitive trait based on an individual's passion for a particular long-term goal or end-state, coupled with a powerful motivation to achieve their goal. We deploy grit at work when we seek delayed rather than immediate gratification, and forego short-term temptations to pursue distant goals that have long-term beneficial outcomes.

Gritty people tend to persevere, self-regulate and push themselves toward success. Duckworth and Seligman[287] found that the correlation between self-discipline (the capacity to persist) and achievement was twice as large as the correlation between IQ and achievement. Grittiness equates to two core dimensions: passion – staying fast/holding onto the goals that you set (the 'will') and perseverance (the 'way'). The neural processes underlying grit show an association between grit and activity dorsal medial PFC, which is involved in self-regulation, planning, goal setting and maintenance, and counterfactual thinking for reflecting on past failures. Behavioural markers of grit include: "I finish whatever I begin", and "I have overcome setbacks to conquer an important challenge". Grit can be learned as its part of a Growth Mindset.

Neuro-Insight: Reflection on Purpose

The source for developing purpose and aligned goals is quality questions, for example, instead of asking: "Why am I hear?" or "What shall I do with my life", ask "What is our/my service to the world?" or "What should our company/my life do for others?"

To be clear, deliberate and inspired by this purpose is where the motivational value lies.

NOTES

_____

_____

_____

_____

_____

_____

_____

_____

#Neuro-Hacks for stratifying purpose and goals

- Cultivate **insight creation as a thinking process:** the acronym ARIA[288] is very useful – it stands for ARIA which stands for **A**wareness of a Dilemma, **R**eflection on current realities and future possibilities, Insight (or AHA moment) and **A**ction.

- **Define your purpose – at the organisational, team and individual level.** The vitality of organisational and individual life is a function of the clarity of the vision.

- **Frame goals** into **social causes**. This talks to what Maslow called self-transcendence; in the brain it equates to oxytocin being released in the ventral striatum. _#giving is its own reward._

- **Focus on the vision and purpose** instead of the emotional charge surrounding a goal. These are also called goal-oriented thought processes. Focusing helps to identify the type of thinking we are doing at any moment and provides an opportunity to then choose where to put our focus. It is a model that supports neuroplasticity in that it enables us to keep on going back to vision based thinking, the goal and the outcome. Building an intrinsic innovation mindset #failfast.

- **Create congruence in your goals.** When our goals are aligned with our highest values or purpose, we become inspired from within. This is congruence – a highly rewarding state where all your actions lead towards what really matters. In this regard, a powerful reflective question is, "How does whatever happens help you achieve your purpose?"

- **Truly democratise** goal setting by practicing a **deliberative goal setting style**. This is about seeking a shared sense of direction, applying good reason and rationale (which is practical and meaningful) to setting goals, and doing it with respect and warmth using an inclusive approach where each others' views are treated with respect. This results in goal commitment where you are willing to take both pain and pleasure to achieve the goal.[289]

- **Create goal clarity.** This means building attention density for the goal set via EFT (Episodic Future Thinking) and doing it in a state of grace, so that you do not give into volatility (emotional charge) when the pain is starting to exceed the pleasure of achieving the goal.

- **Forget motivation and develop discipline instead. Be a master of persistence and do what it takes.** Developing the discipline to do something you need to without needing to feel motivated is empowering. This is grit, which is passion and perseverance for long term goals, similar to conscientiousness.

Conclusion

Our brains and our lives are mirror images of each other. This is because the brain is the core system that underpins all our emotions, thinking, feelings and how well we regulate them to achieve our purpose, goals and our sense of meaning. Successful goal pursuit requires much more than simply setting goals. First, one needs to participatively create a desired end state (or purpose) then a clear line of sight through goal specificity. Lastly, one needs to take the right repeated actions to build new brain wiring.

7

Putting it all together

We measure much of what we do at work through financial capital. In a highly competitive global marketplace, each form of capital – financial, human, social and psychological – can serve as a rich source of business advantage. For this reason, leveraging capital is an ongoing imperative of business leadership, but doing so is hard because it often requires changing the way hundreds of people think and behave. And no-one likes change.

Everything discussed in this book is intended to simplify complex ideas and concepts from the social cognitive and affective neuroscience literature into meaningful strategies and practical action steps.

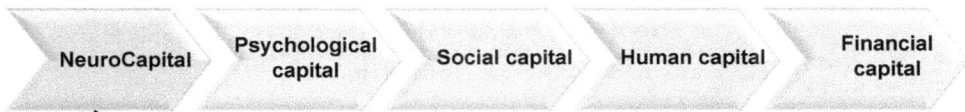

1. Stratification of goals
2. Social connectivity
3. Simple fluency and regulation of emotions
4. States of mood
5. Stillness of mind
6. Sleep health
7. Stress resilient

To get the most out of this book, it will be useful to revisit some things:

Firstly, the SCANS model – the foundational concepts for NeuroCapital in the world of work as shown in Figure 28.

Safety is first	**The Brain "wires to survive" not thrive:** Our brains distinguish the threat and reward content of every single experience we encounter, mostly subconsciously, and we behave in accordance with the brain's assessment.	**"Learn to ride the T-Rex"**
Conscious thinking is overrated	**The Brain uses two modes of processing the subconscious and the conscious:** Reflexive and Reflective modes of memory systems that feedforward and feedback emotional "action tendencies". The reflexive mode is expressed as implicit memories or autopilot tendencies.	**"Primal patterns are all powerful"**
Attention is an open system	**The Brain is an open system:** The brain is a neural network that does not exist in isolation. It is in close symbiosis with its environment. When the environment is compromised, the neural system becomes compromised as it wires itself to survive and disrupts effective neural proliferation.	**"Energy flows where attention goes"**
Neuroplasticity	**The Brain is plastic – Neuroplasticity:** The brain has the ability to rewire itself based on where we focus attention. Genes dictate the overall architecture of the brain but the structure is dynamic, continuing to regenerate cells and changes throughout our lives.	**"Neurons that fire together wire together"**
Scaling the brain into networks	**The Brain is social first:** We are hardwired for social connectivity, and this need is as rudimentary as food and water for survival. Thus, our social cognition is ancient and by design.	**"Social wiring is ancient and by design"**

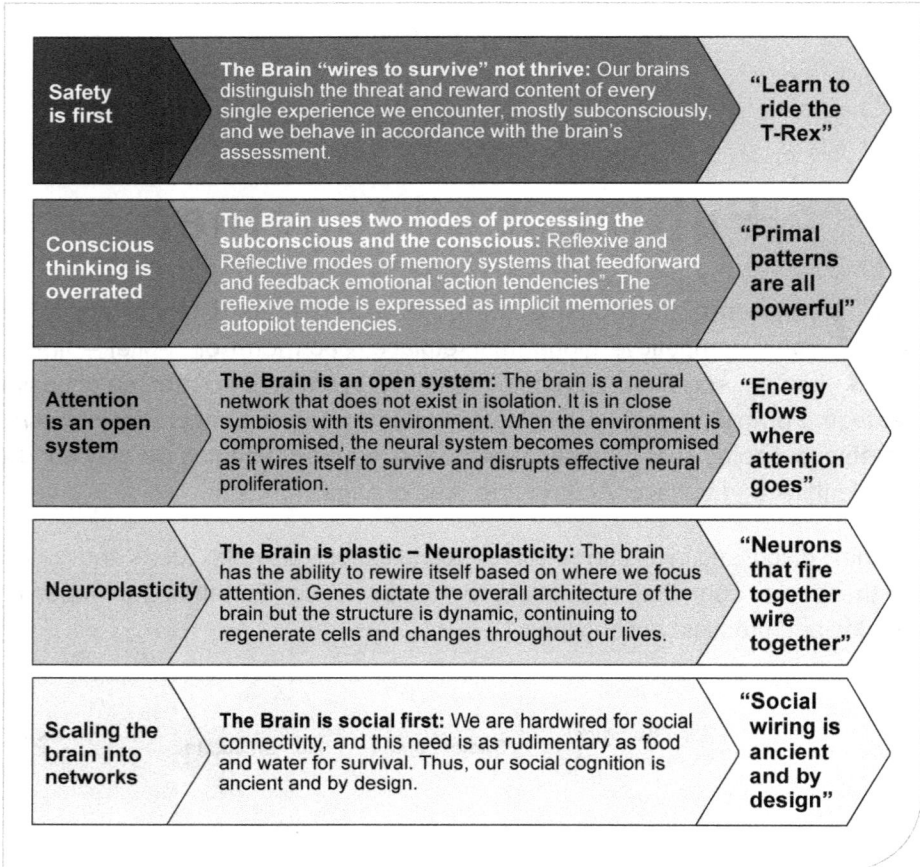

Figure 28: The SCANS model

Secondly, the seven strategies for developing your NeuroCapital proficiencies.

Use the seven strategies (set out in Figure 29) to learn and teach yourself, team members, or coaching clients about your/their brain, and how to regulate and train your/their brain to move from "fight" or "flight" to "flourish".

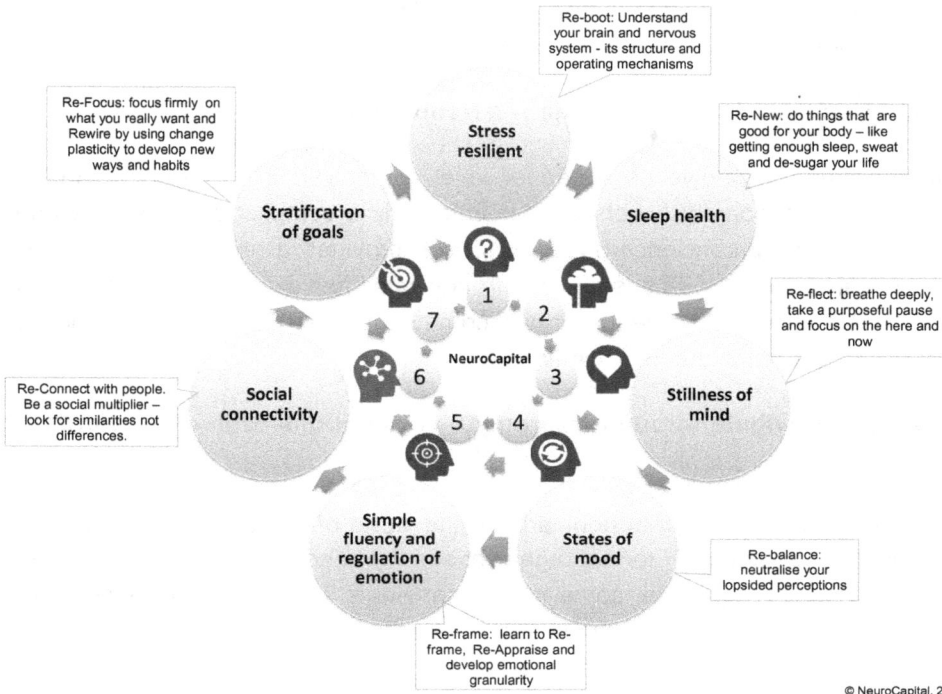

Figure 29: The seven strategies for developing your NeuroCapital

The NeuroCapital strategies discussed have differential and beneficial effects that complement each other, providing a well-balanced "repertoire" for optimal functioning and well-being.

1. **Re-boot:** Understand your brain and nervous system, including their structures and foundational concepts.

2. **Re-new:** Do things that are good for your body, like getting enough sleep, sweating and de-sugaring your life.

3. **Re-flect:** Breathe deeply, take a purposeful pause and focus on the here and now.

4. **Re-balance:** Neutralise your lopsided perceptions to stabilise your mood.

5. **Re-frame:** Transform your perceived negatives into positives (use your discomfort for growth). Name it to tame it. Re-appraise. Develop emotional granularity.

6. **Re-connect** with people. Be a social multiplier – look for similarities, not differences.

7.   **Re-focus**: Focus firmly on what you really want and how this can help others; rewire by using change plasticity to develop new ways and habits.

Thirdly, cultivate a new language – a Neuropedia.

Neuroscience may give us a deeper level of analysis in the study of our work behaviour, but workplace behaviour is inherently a complex process. The applied organisational neuroscience-based approach provides a neuroscientific language for mental experience. To ensure that this new and often disruptive language enables the transference of knowledge, it is recommended that you delve into the useful definitions set out in the neuropedia in Appendix E.

Fourthly, Navigate through Neuromyth Conceptions or in other words - get the Data!

Neuroscience is growing in popularity in the world of work. To navigate this new science in the context of formal organisations we need to be aware of and debunk Neuromyths. For example, some myths; cortisol is THE stress hormone, we only use 10% of our brain, the Amygdala is the fear factory of the brain, there are basic emotions which are universal and that have unique biological mechanisms (to name a few Neuromyth conceptions). The way to untangle these mythconceptions and confusion is to look for evidence-based research (ideally meta-analysis studies) and to adopt a practitioner scientist stance in your conversations, coaching, consulting, leadership and strategizing practices. There are many references in this book that can serve as a navigation starting point

Lastly, Start with YOU – There can be no team or organisational transformation without personal transformation. Then implement systems or approaches that scale up to create a collective shift.

Final reflection

Our world at work is complex and ever-changing. While evidence from neuroscience to inform behaviour at work may not reduce the uncertainty and complexity, it can perhaps assist in providing a framework in which to operate with more insight in this space.

Appendix A: Thinking frameworks and models

Various thinking frameworks and models exist to explain brain functioning. I focus on those that can be applied to the manifestations of the brain in the workplace and their application to leadership behaviour, namely the triune brain model, large-scale network models, and the integrative neuroscience model.

1. The Triune Brain Model

The neuroscientist Paul MacLean proposed a model of the development of the vertebrate forebrain – the *triune brain* – as shown below in Figure 30. The triune brain model offers a broad theory of brain evolution, or more precisely, how the collection of neurons forms various structures and regions. MacLean called the evolutionary layers the reptilian, the paleomammalian and the neomammalian brains, which parallel the structures of the brain stem, limbic system and neocortex, with each layer having increasingly complex functionality.

According to the triune model, lower order, more primitive processes tend to override more advanced cognitive processes. This is because the brain strives to conserve energy and reverts to established patterns and habits engrained in those 'older' regions. Although it is easy to see the three layers of the brain, from brain stem, to limbic system, to the cerebral cortex, as separate functional areas processing living functions, emotions and higher functions, these meta-regions are all linked together and process information in parallel and with the help of each other.

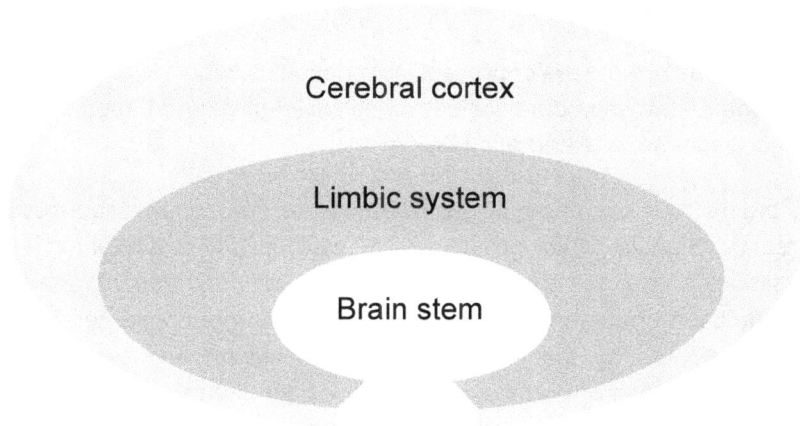

Figure 30: The Triune Brain Model (Source: MacLean)[290]

153

2. The Integrate Model

The Integrate Model, which proposes a dynamic continuum of brain organisation, brings together a number of theories across disciplines and scales.[291] The Emotional Brain Theory of LeDoux[292], which stipulates the biological underpinnings of emotion and memory, especially brain mechanisms related to fear and anxiety, is a cornerstone of the Integrate Model. According to LeDoux's theory, there are three implications for behaviour: the speed of activation, the sequence of activation, and the endocrine/neural system interactions. The firing of the limbic system seems to occur within eight milliseconds of a primary cognition, but it takes almost 40 milliseconds for that same circuit to appear in the neocortex for interpretation and conceptualisation.[293, 294]

With this timing, our emotions are determining cognitive interpretation more than previously recognised. Once primary cognitions have occurred, secondary cognitions allow for the neocortical events (i.e. reframing) to drive subsequent limbic or emotional labelling. Our unconscious emotional states are arousing emotions in those with whom we interact before we or they know it. And it spreads from these interactions to others.

The integrative neuroscience approach brings together crucial organising principles across scales of brain function. This has resulted in the integrate or 124 Model, which outlines the brain's core motivations and key functional modes. The 124 Model is a framework for explaining brain performance:

- The brain's core organising principle is safety first. Thus, the brain wires itself to survive rather than to thrive.

- The brain has two modes of processing: conscious, which is rational, verbal and detail orientated; and nonconscious, which is intuitive and based on awareness of, and response to, external cues.

- The brain's four key underlying processes are emotions (responses to threat or reward signals), thinking (focusing, memorising and planning), feelings (physiological changes in heart rate, breathing and perspiration), and successful self-regulation of these functions. All these highly interconnected brain circuits are underpinned by the exquisite timing of electrochemical activity. These include the release of noradrenaline for the fight-flight response, dopamine for reward cues, serotonin for enhancing one's mood, and oxytocin for bonding.

In order to manage information and sensory demands, the human brain operates by using parallel distributed processors, which means that many operations are going on simultaneously, gathering together into larger networks at a later stage in processing. A core principle is a motivation to 'Minimise Danger and Maximise Reward'. This motivation helps a leader to deal with immediate threats, but also drives the search for rewards over longer timescales – from nourishment, through social connectivity, to purpose in life. The core 'Minimise Danger and Maximise Reward' principle continually organises the fundamental brain processes of emotion, thinking, feeling and self-regulation.

The 124 Model seen in Figure 31 below highlights the timing of the brain's key processes. Many emotional reactions, for instance, occur within a fifth of a second, without any conscious awareness.

The emotion-feeling-thinking and self-regulation processes that are unique to each person are shaped by both genetic disposition and life experiences. **Genes (= nature)** are not destiny, but rather 'disposition'. **Bonding and conditioning experiences (= nurture)** can have a lasting effect from childhood and shape the human brain's ongoing personal experiences. Neural plasticity enables transformative brain changes via the right insights and training, which translate into new behaviours.

The consequences of the 124 Model for leadership behaviour is that the crucial threshold for leadership behavioural change is nonconscious concrete processing, as this mode of processing is automatic and emotionally evocative. Brain effectiveness is primarily determined by how well leaders can train and align their nonconscious and conscious modes of processing.

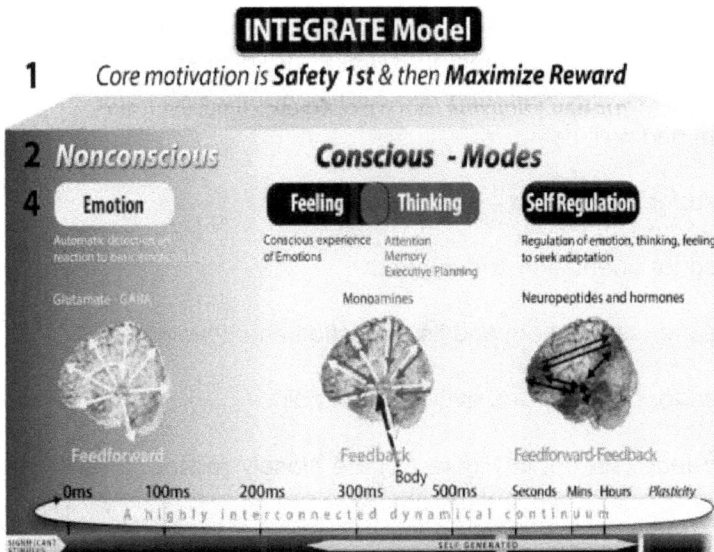

Figure 31: The 124 Model (Source: Gordon)[295]

3. The Neuro-psychotherapy Model

Groundbreaking research on the Aplysia California slug by molecular neuroscientist Eric Kandel, a Nobel Laureate, transformed the model of contemporary neuroscience from an electrochemical paradigm to decode neural activation to that of the brain as a neural network with unknown potential to be altered (neuroplasticity) at a molecular and cellular level. The pivotal publication, *A new intellectual framework for psychiatry*, captured Kandel's significant scientific revolution in molecular neuroscience. This engendered changes from the bygone functional approaches to the new dynamic view of the brain. Kandel's work pioneered new propositions regarding the link between the brain and the environment, genes as communication agents, the interconnectivity between nature and nurture, the effect of enriched social environments on brain development, and even the role of talking therapies to facilitate changes to the brain.[296, 297, 298]

A Basic Needs Model

From an evolutionary perspective, human functions have developed so as to use the environment to its best and allow the reproduction and development of the species – its survival and growth. This raises the question of what the optimal conditions for human development are precisely. These are firstly the physiological basic needs that drive our physical survival: hunger, thirst and sleep. These needs have been extensively researched in many forms.

Our **psychological needs** are represented by many different approaches and interpretations. Grawe[299] defined four basic needs that are present among all humans, the violation or enduring non-fulfillment of which leads to impairments in mental health and well-being.

- The need for attachment – belonging.

- The need for orientation and control.

- The need for self-esteem and its protection and development.

- The need for pleasure and avoidance of pain.

These four needs (set out in Figure 32) are closely related to each other, and the satisfaction of one will influence the others.

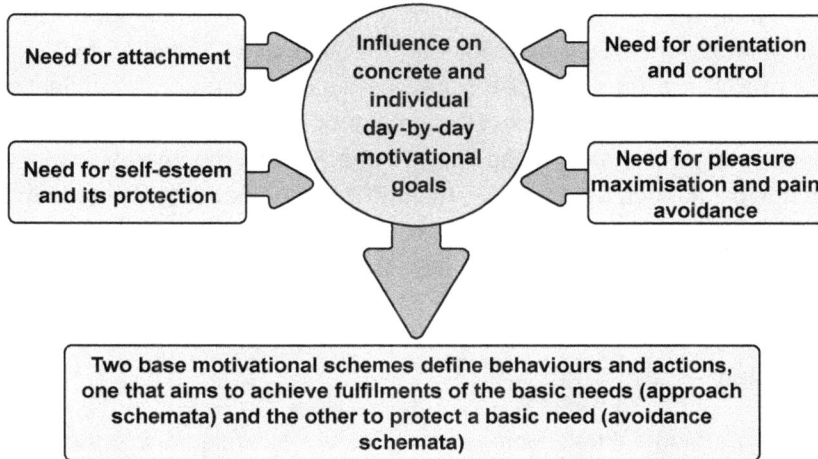

Figure 32: The four basic needs/factors that influence human behaviour (Source: Ghadiri, Habermacher & Peters)[300]

Summarising the basic needs

These four basic needs lie at the emotional heart of human beings. They differ slightly but are also related to the survival instincts which lie at a primitive and deeper level in the brain. We can think of the survival instincts as the brain stem and the emotional needs as the next level, the limbic level. As per the triune/three-layer model of the brain, processes are interlinked so our survival needs will colour our emotional needs, and our emotional needs may colour each other. These may also work in separation. The reward centre, which is directly linked to pleasure, is also stimulated by primary rewards, e.g. our survival instincts, food and sex. Fulfillment of one or more basic needs will also stimulate reward and hence colour the basic need for pleasure. For example, being complimented will influence our basic need for self-esteem, but it will also be a rewarding experience, increasing our pleasure. The same applies to attachment and orientation and control – these can all stimulate reward and hence pleasure. If these needs are not fulfilled or in balance, then a person cannot be in harmony and this will manifest itself in disruptive behaviour and well-being, and in more extreme cases lead to psychological disorders.[301]

4. The Neuro-Triangles Model

The Triangles Model can be viewed as a practical, empowering application supporting the enhancement of wellness and performance, and thus, personal efficiency and gratification.[302]

The Triangles Model and its diagnostic application were developed using predominantly psycho-neuroimmunology (PNI) as well as logotherapy.[303] The main interests of PNI are the interactions between the nervous and immune systems, specifically the relationships between mental processes, behaviour and health. PNI was pioneered by Robert Ader who showed conclusively that manipulating cognitive function could influence aspects of immunity.[304] In longstanding PNI research, it has been demonstrated that there is a bi-directional communication flow between the brain and the immune system, that is, from the brain to the immune system and also from the immune system to the brain.[305] Hormones, neuropeptides and cytokines facilitate this bi-directional communication between the nervous, endocrine and immune systems.[306]

In the Triangles Model, identifiable mind states are associated with the secretion of neurotransmitters and neuropeptides, which then circulate in the bloodstream and impact upon body metabolism. In this way (see Figure 33), mind states influence wellness and performance in a profound way.[307]

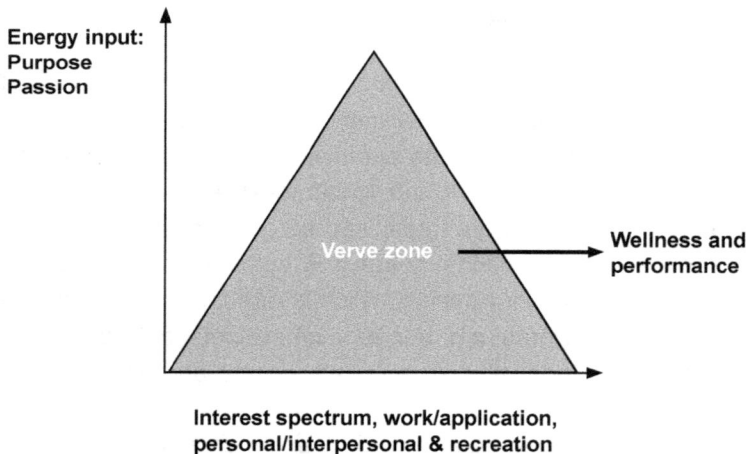

Figure 33: The energy-integration curve source (Source: Weinberg)[308]

The nature-nurture heritage gives rise to our worldview or subjective reality. If this heritage was comprehensive and free of deprivation, then a person could interact with the external environment, as they would see things as they really are and consequently make accurate decisions.

To learn more about the Neuro-Triangles model visit: www.PNINET.com

5. The Neurozone Model

The Neurozone Model and diagnostic of brain performance were developed by the neurologist, Dr Etienne van der Walt.[309] The neurozone diagnostic offers insight into the brain's constituents for innovative performance. The model consists of **real brain structures around a hypothetical axis from basic to most sophisticated, to support the functional need to solve a variety of daily problems**. This is a vastly unconscious process that keeps us alive and helps us thrive. The model clusters 10 key drivers[310] into four performance conditions for individuals and organisations to thrive: *resilience, learning, leadership* and *innovation*. These are referred to as brain performance conditions because they are natural outcomes of actively enhancing the 10 neurozone drivers of brain performance. In Table 9 below, I identify specific drivers that most directly optimise each of the four brain performance conditions for thriving.

Table 9: The brain performance conditions and key drivers

Brain Performance Conditions for Thriving	Optimising Brain Performance Drivers
Resilience	Exercise – Sleep/Wake Cycle – Value Tagging – Silencing the Mind
Learning Capacity	Exercise – Sleep/Wake Cycle – Nutrition – Learning (as a driver) – Silencing the Mind
Self-Leadership	Value Tagging – Goal-Directedness – Collective Creativity – Silencing the Mind
Innovation	Learning – Abstraction – Execute Function – Silencing the Mind

Summary of the brain performance conditions[311]

- **Resilience** is the capacity of the brain/body system to withstand the challenges that threaten its stability. Resilience is an active process and is adaptive, for example, the brain actively and continuously adapts at a molecular and neuronal level in the presence of a stressor. This adaptive capability determines its resiliency. Effectively enhanced resilience prevents an implosion of the brain/ body system and sets individuals up for optimal performance.

- **Learning capacity** can be viewed as the capacity of the brain/body system to register, store and consolidate information as insightful knowledge, and to be able to retrieve this useful information when needed. It does this by forming memories; the hippocampus is the key brain structure in this process. In fact, without the hippocampus the brain cannot form memory. There is more potential for neurogenesis (new brain cell formation) in the hippocampus than in any other part of the brain.[312]

- **Self-leadership** is defined as the ability to have a clear vision of the goal that needs to be achieved, to accurately calculate the strengths and resources available to achieve the goal, to provide adequate energy to drive the process, and to effectively integrate learning from the process. In this sense, self-leadership is an imperative for leadership of the collective.[313]

- **Innovation capacity** is the collective capacity of the brain/body system to solve problems and fashion novel products. The brain has an astonishingly complex and sophisticated capability to form "realistic scenarios of the possible" – this lies at the heart of innovation. From a neuroscience perspective, creativity is the highest form of thinking and always precedes innovation. And while the capacity to innovate is fuelled by continuous acquisition of knowledge and skills, the spark of ignition is the willingness to embrace novelty and diverse perspectives.[314]

To learn more about Neurozone visit: www.Neurozone.com

6.   The SCARF Model®

The SCARF Model[315] is based on social, cognitive neuroscience and represents the social elements that activate either primary rewards, or primary threat circuitries, in the brain. Reward states are associated with more cognitive resources[316], while a threat state is associated with impulsive, instinctive and avoidance responses such as sadness, anxiety, anger and avoidance.[317]

The SCARF Model involves five domains of human social motivation: Status, Certainty, Autonomy, Relatedness and Fairness:

- Status is about relative importance to others.

- Certainty concerns being able to predict the future.

- Autonomy provides a sense of control over events.

- Relatedness is a sense of safety with others – of friend rather than foe.

- Fairness is a perception of fair exchanges between people.

Organisational examples:

1. Status is about relative importance to others. Given that our brains are socially wired, we are constantly scanning to see where we sit in our social or professional group.

Status threats	Status rewards
Excluding team members from joint decision making processes	Calling on various team members for their subject matter expertise

2. Triggering a threat response.

Certainty threats	Certainty rewards
Shifting goal posts re focus and priorities	Providing future expectations
Mixed messages	Providing more frequent open communication

3. Autonomy provides a sense of control over events.

Autonomy threats	Autonomy rewards
Rule bound with strict command and control decision making	Increase the amount of discretion that people have at work to make decisions

4. Relatedness is a sense of safety with others – of friend rather than foe.

Relatedness threats	Relatedness rewards
Allowing minor aggression, i.e. name calling	Identifying splitting and pairing (cliques) dynamics in a team Work towards diversity and inclusion

5. Fairness is a perception of fair exchanges between people.

Fairness threats	Fairness rewards
Perception of bias – employees put in extra hours without fair reward or incentives	Transparent on who qualifies for incentives, rewards etc.

To conclude, the SCARF Model provides a means of alerting (bringing to conscious awareness) employees' core social motivators (which they might not be aware of themselves). The process starts by reducing the threats inherent in the workplace. It enables employees and management to be more aware of their own needs and the social conditions that accelerate performance and engagement. Understanding that these five domains are primary needs helps individuals and leaders better navigate the social world in the workplace.

7. The S.A.F.E.T.Y.™ Model

Overview: As society has evolved, our brain has become acutely sensitive to the psychological threats you experience in your social interactions – things like attitudes, behaviors, and perceived motivations. So how do you identify and manage the triggers that threaten your psychological safety and hijack your brain? The S.A.F.E.T.Y model[318] provides a classification system of essential social drivers that ensure our social cognition is optimised for flourishing. Based on the very latest neuroscience research, this model describes some of the most important social motivators of human behaviour (as set out below).

Psychological the S.A.F.E.T.Y model explains how to implement it in your life and your workplace, to reap the benefits of increased productivity and personal well-being.

S.A.F.E.T.Y is a guide to navigating key scenarios for maximum benefit, from onboarding and corporate restructures to brainstorming sessions and reward programmes.

S.A.F.E.T.Y.™ MODEL

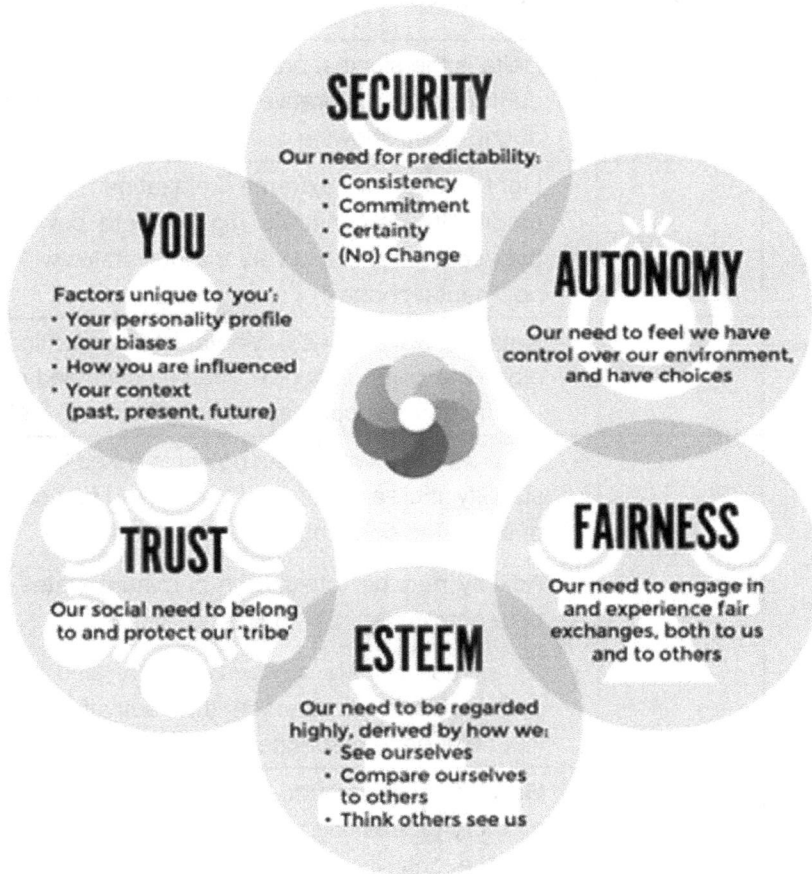

SECURITY
Our need for predictability:
- Consistency
- Commitment
- Certainty
- (No) Change

YOU
Factors unique to 'you':
- Your personality profile
- Your biases
- How you are influenced
- Your context
  (past, present, future)

AUTONOMY
Our need to feel we have control over our environment, and have choices

TRUST
Our social need to belong to and protect our 'tribe'

ESTEEM
Our need to be regarded highly, derived by how we:
- See ourselves
- Compare ourselves to others
- Think others see us

FAIRNESS
Our need to engage in and experience fair exchanges, both to us and to others

©2018 The Academy of Brain-based Leadership

Figure 34: S.A.F.E.T.Y.™ Model[319]

Appendix B: Categories of brainwaves and mental states

Brain Wave	Frequency	Mental States
Delta	0 – 3Hz	• Delta is the slowest brainwave type, and is dominant during deep, restorative sleep. A high level of delta during sleep shows a deeper, more stable sleep state
Theta	3 – 8Hz	• Light sleep or extreme relaxation. Experienced meditators demonstrate higher theta power during their peak practice, in very tranquil states with almost no "mental chatter"
Alpha	8 – 12Hz	• Relaxed wakefulness. Mental states dominated by alpha are commonly described as calm and pleasant – sometimes accompanied by a "floating" feeling
Beta	12 – 27Hz	• Wide awake. Higher beta power is linked to emotional stability, increased energy levels, and focused states of extended concentration • Anxiety may be reflected in dominant states of high beta amplitudes: 21-25 Hz • An excess of beta brainwaves may also indicate a narrow focus of attention that activates the fight-flight-or-freeze response
Gamma	27+ Hz	• Gamma is associated with the formation of ideas, memory processing and learning. Humans and apes are the only living creatures that produce gamma waves

Appendix C: Neurotransmitters and their functions

The brain is a biological electrical circuit of incredible complexity which is driven by biological electrical processes that are stimulated by a host of different chemical substances. Below I examine the four key transmitters in the brain that influence broad circuits and key functions, namely: acetylcholine, serotonin, dopamine and noradrenaline. These transmitters work in large networks of neurons and regions.

Neurochemicals

The **serotonin** system plays a key role in mood creation, particularly in fear and aggression. Serotonin is also produced in the gut (90% is actually produced in the digestive tracts), thus it gauges food availability (and is thus responsible for irritability and anger, particularly amongst men, when hungry). It also has other wide-reaching effects, varying from influencing cardiovascular health to gauging social situations, mating behaviours and social status.

The **dopamine** system is an extensive network that influences key areas in the brain and generates feelings such as euphoria. The dopamine system runs from the limbic system to the prefrontal areas of the brain where our executive functions sit. It is therefore an important element of attention; attention deficit hypertensive disorder (ADHD) is thought to have roots in a lack of dopamine. Dopamine also serves as our reward drug, inducing feelings of happiness and euphoria. On the negative side, it plays a role in compulsion. In combination with oxytocin, it produces powerful bonding feelings.

Noradrenaline (or norepinephrine) is produced from dopamine and is responsible for the fight-or-flight response. Its effect includes increased heart rate, increased oxygen supply to the brain and muscles, and release of glucose in the bloodstream. It is therefore a key stress response chemical. It also plays a role in attention, particularly with reference to realistic shift prediction. As with dopamine, lowered levels of this neurochemical are implicated in ADHD. On a positive note, there are also times when adrenaline is involved, like when we are falling in love or when we are in the zone with a work assignment.

Oxytocin is a neuromodulator that is released from the pituitary gland and is most well-known for its role in trust. It is associated with the experience of empathy (both expressed and received), at it has been shown to diminish the secretion of cortisol, reduce blood pressure, increase pain thresholds, have an anti-anxiety effect (through suppressing the activity of the limbic system), and stimulate positive social interactions.

There are currently over 50 substances that we know of that operate as transmitters in the brain. Drugs and medication can magnify, decrease or inhibit their influences.

Appendix D: Bias and mental short cuts

1. The false consensus bias

False consensus bias is an attributional type of cognitive bias where we believe that people like us hold the same beliefs as us.

We overestimate the extent to which our opinions, beliefs, preferences, values and habits are normal and typical of those of others (i.e. that others think the same way that we do).

This cognitive bias tends to lead to the perception of a consensus that does not exist, i.e. a false consensus.

Dialling it down:

* Down-regulate stress and cognitive load.

* Know that perception is projection (the tendency to assume that others share our cognitive and affective states).

* Maximise face-to-face interactions and cultivate counter voices – allow "yes but thinkers".

2. The confirmation bias – "Reflexive reasoning"

The tendency to search for, interpret, favour and recall information in a way that confirms one's pre-existing beliefs or hypotheses.

Our expectations are shaped by our previous experiences, and our automatic reactions make it possible for us to live in a way that is functional and safe.

The effect is stronger for desired outcomes, emotionally charged issues, and for deeply entrenched beliefs.

A recruitment and selection example could be: "We only employ graduates from reputable universities."

Dialling it down:

* Have deliberate exchanges with employees who are perceived as "different", e.g. have lunch dates to learn more about dissimilar others.

- Learning more about someone is likely to reshape our assumptions about them based on more pertinent information, and we are thereby more likely to recast our daily perceptions of their actions.

A virtuous confirmation bias gratitude in the brain: Driving action bias. Feelings of gratitude directly activate the neurotransmitter, dopamine. Dopamine = the "feel good" neurotransmitter. Gratitude can have a powerful impact on your life because it engages your brain in a virtuous cycle; your brain only has so much power to focus its attention.

Gratitude takes practice like any other skill. Try thinking of three things every day that you are grateful for.

3.   The insider bias – "Beware of strangers"

Our brains are wired to recognise and respond to threats to our survival, and to immediately recognise subtle cues regarding whether another person is part of what we regard as our "in-group", or a potentially hazardous "out-group".

Dialling it down:

- Widen your "in-group" circle by creating new markers of membership such as clothing with the company logo, team pictures or invitations to corporate events.

- Create shared experiences and goals (aligning behind a common course breaks down perceived differences).

- Compensate for implicit preferences, for example, engage dissimilar others to truly understand them better or watch TV shows to learn about other cultures (look for similarities, not differences).

4.   The attribution bias – "It wasn't me"

Decades of psychological research have established that when there is a problem, we tend to explain our own actions in terms of circumstances and attribute the actions of others to character flaws; the reverse is true when the results are positive. "Sorry I'm late. The traffic was terrible." "He is always late. Being on time is an occupational hazard."

Dialling it down:

- Question character judgements, particularly regarding those who might not be perceived as "insiders", by asking about mitigating circumstances. "Is he frequently late or was this an exception? Did he have a prior meeting?"

5.   The over confidence bias – "All or nothing thinking"

We have greater subjective confidence in our judgements than an objective assessment would warrant. We also tend to overestimate our own performance relative to that of others.

Dialling it down:

- Test confident assertions, both your own and those of others, for signs that they are grounded in solid evidence.

- Systematically incorporate multiple perspectives into processes such as succession planning to better ensure that assertions are examined from various points of view using balanced sources of data. For example, have a triangulation of data points such as structured interviews with diverse panels, diagnostics, and biodata. This will ensure less bias and more objectivity.

Appendix E: Neuropedia – some helpful definitions

Adrenaline: Best-known for its role in the fight-or-flight response.

Allostatic load: A range of markers of stress, including cortisol and adrenaline levels in the blood, as well as immune system activity and blood pressure.

ACC: A salient or control node in the brain responsible for conflict or error detection.

Amygdala: A small brain region that is part of the limbic system, which activates based on the strength of an emotional or motivational response.

Basal ganglia: Key to storing routines, repetitive behaviours and thoughts – the home of habits.

Blood-Brain Barrier (BBB): A barrier surrounding the brain that prevents some materials from entering while allowing others in (e.g. glucose), thereby helping to maintain a constant environment for the brain.

CEN: Brain network responsible for high-level cognitive functions, notably the control of attention and working memory.

Cortisol: Notably named the 'stress hormone' because it is released in response to stress.

DMN: A large-scale network of brain areas roughly in the middle of the brain, including the medial prefrontal cortex. It forms a unified system of self-related cognitive activity, including autobiographical, self-monitoring, and social functions. The DMN is typically deactivated during stimulus-driven cognitive processing. It activates when you are not doing much else, and when you think about yourself and other people.

Dopamine: One of the two main neurotransmitters involved in stabilising circuits in the prefrontal cortex (norepinephrine is the other). Dopamine is connected to feeling interested in something and is important for tasks that involve learning new things.

Dorsolateral prefrontal cortex (DLPFC or DL-PFC): An area in the prefrontal cortex of the brain of humans and non-human primates. It is one of the most recently derived parts of the human brain. It undergoes a prolonged period of maturation, which lasts until adulthood. The DLPFC is not an anatomical structure, but rather a functional one. It lies in the middle frontal gyrus of humans.

Endocrine: Pertains to the glands that secrete hormones into the bloodstream, which circulate around the body. This system includes estrogen, testosterone, growth hormones, adrenalin and insulin, amongst others. Glands include the thyroid, parathyroid, pancreas, pineal and gonads.

Epigenetics: An emerging science looking at how environmental effects might influence the way genes are expressed.

Functional magnetic resonance imaging (fMRI): A form of non-invasive neuroimaging based on blood oxygen level-dependent signals in the brain in vivo.

GABA and glutamate: The king and queen of neurotransmitters.

Hippocampus: Has roles in the consolidation of information from short- to long-term memory.

Hypothalamus: Responsible for various metabolic processes and synthesising and secreting neurohormones.

Insular cortex: Enables a degree of awareness and is involved in our ability to be 'in tune' with ourselves.

Limbic system: A region in the centre of the brain important for experiencing emotions, memories and motivations; includes the amygdala, insula, hippocampus and orbital frontal cortex.

Mindfulness: A form of meditation, derived from Buddhism, involving the practice of focusing attention on what you are experiencing, in the present, non-judgementally.

Mirror neurons: Neurons in the brain that help us directly experience other people's intentions, motivations and emotions, by feeling the same way ourselves.

Myelination: The process by which a fatty layer, called myelin, accumulates around nerve cells. Myelination enables nerve cells to transmit information faster and allows for a more complex brain process.

Neuroplasticity: The study of change in the brain, both moment to moment and in the long-term.

Oxytocin: Involved in social behaviour, increasing trust, decreasing fear, increasing generosity and improving cognitive functions.

Posterior parietal cortex (PCC): The portion of parietal neocortex posterior to the primary somatosensory cortex. The PCC plays an important role in planned movements, spatial reasoning and attention. Damage to the posterior parietal cortex can produce a variety of sensorimotor deficits, including deficits in the perception and memory of spatial relationships, inaccurate reaching and grasping, the control of eye movement, and inattention. The two most striking consequences of PPC damage are apraxia and hemispatial neglect.

Prefrontal cortex: A section of the outer layer of the brain, behind the forehead, which is involved in many types of executive functioning, planning and coordinating the rest of the brain.

SN: A large-scale brain network involved in detecting and orientating towards salient external stimuli and internal events.

Serotonin: Important for mood regulation, appetite, sleep, and memory and learning.

Striatum: Involved in pleasure, reward, motivation, reinforcements learnings, fear and impulsivity.

Ventrolateral prefrontal cortex (VLPFC): A region of the prefrontal cortex, beneath the right and left temples, that is important for all types of braking functions, including stopping physical movement and inhibiting emotions or thoughts.

Working memory: A term used to denote that part of memory which acts as temporary storage – a 'holding tank' for information that we need to have or manipulate to perform a current activity. An example is holding a phone number in your mind while you dial it. Working memory has a very limited capacity.

NOTES

_____

_____

_____

_____

_____

_____

_____

Endnotes

1   Rilke, R. M., & Herter, N. M. D. (2004). *Letters to a young poet*. Retrieved from: https://en.wikipedia.org/wiki/Letters_to_a_Young_Poet

2   American Psychological Association. (2018). *APA Dictionary of Psychology*. Retrieved from: https://dictionary.apa.org/pdf

3   Feynman, R.P. & Robbins, J. (2005). *The Pleasure of Finding Things Out*. New York: Basic Books, p292.

4   Ibid.

5   Artley, J. (2018). *How to be a leader in the Fourth Industrial Revolution – Notes by Professor Klaus Schwab*. Retrieved from: https://www.weforum.org/agenda/2018/01/how-to-be-a-leader-in-the-fourth-industrial-revolution/

6   Edmondson, A.C. (2018). *The Fearless Organization: Creating Psychological Safety in the Workplace for Learning, Innovation, and Growth*. Hoboken, New Jersey: John Wiley, p21.

7   Nicholson, N. (1998). How hardwired is human behavior? *Harvard Business Review, 76(4), 134-147*, p135.

8   Espinosa, K. J. P., & Caro, J. D. L. (2011). A real-time web-based Delphi study on ICT integration framework in Basic Education. In *international conference on telecommunication technology and applications*.

9   Gordon, E., Palmer, D. M., Liu, H., Rekshan, W., & Devarney, S. (2013). Online Cognitive Brain Training Associated With Measurable Improvements in Cognition and Emotional Well-Being. *Technology and Innovation, 15*(April 2012), 53–62. https://doi.org/10.3727/194982413X13608676060574

10   Higgins, D. (2013). *Reflective Learning in Management, Development and Education*. Abingdon: Routledge.

11   BrainyQuote. (n.d.). *Albert Einstein Quotes*. Retrieved from: https://www.brainyquote.com/quotes/albert_einstein_103652

12   Lindebaum, D. & Zundel, M. (2013). Not quite a revolution: Scrutinizing organizational neuroscience in leadership studies. *Human Relations, 66*, 857-877.

13   Ibid.

14   van Ommen, C., & van Deventer, V. (2016). Negotiating neuroscience: LeDoux's "dramatic ensemble." *Theory and Psychology, 26*(5), 572–590. https://doi.org/10.1177/0959354316659555.

15   Sapolsky, R. M. (2017). *Behave: The biology of humans at our best and worst*. New York, NY: Penguin, p386.

16   Ward, J. (2016). *The student's guide to social neuroscience*. New York, NY: Psychology Press.

17   MacLean, P. D. (1990). *The triune brain in evolution: Role in paleocerebral functions*. Berlin: Springer Science & Business Media.

18   Ibid.

19   Carter, R. (2014). *The human brain book: An illustrated guide to its structure, function, and disorders*. New York, NY: Penguin, p.126.

20   Ibid, p.128.

21   Ibid, p.127.

22   Kriegler, S. (2008). *Getting Out of Here – Now!* (course notes). Pretoria: University of Pretoria.

23 Ibid.
24 Griffiths, P. E. (1997). *What emotions really are: The problem of psychological categories.* Chicago, IL: University of Chicago Press.
25 MacLean, P. D. (1970). *The Triune Brain, Emotion, and Scientific Bias.* In F. O. Schmidt (ed). The Neurosciences Second Study Program. New York: Rockefeller University Press, pp. 336-49.
26 Rutstein, J. S. (2019). *How to Enhance Connection, Happiness, and Ease: The Neuroscience of Self-Regulation.* Retrieved June 12, 2019, from https://www.soundstrue.com/store/brain-change-summit?
27 Porges, S. W., & Dana, D. A. (2018). *Clinical Applications of the Polyvagal Theory: The Emergence of Polyvagal-Informed Therapies (Norton Series on Interpersonal Neurobiology).* New York: WW Norton & Company.
28 Ibid.
29 Ibid.
30 Porges, S. W. (2007). The polyvagal perspective. *Biological Psychology, 74*(2), 116–143.
31 Ibid.
32 Porges, S. W., & Dana, D. A. (2018). *Clinical Applications of the Polyvagal Theory: The Emergence of Polyvagal-Informed Therapies (Norton Series on Interpersonal Neurobiology).* New York: WW Norton & Company.
33 HVMN. (2016). *EEG Measures of Cognition.* Retrieved from: https://hvmn.com/biohacker-guide/cognition/eeg-measures-of-cognition.
34 Zhuang, T., Zhao, H., & Tang, Z. (2009). A Study of Brainwave Entrainment Based on EEG Brain Dynamics. *Computer and Information Science, 2*(2), 80–86.
35 Swart, T., Chisholm, K., & Brown, P. (2015). *Neuroscience for leadership: Harnessing the brain gain advantage.* New York: Palgrave Macmillan.
36 Ghadiri, A., Habermacher, A., & Peters, T. (2013). *Neuroleadership: A journey through the brain for business leaders.* Heidelberg, Germany: Springer Science & Business Media, p50.
37 Ward, J. (2016). *The student's guide to social neuroscience.* New York, NY: Psychology Press.
38 Van der Walt, E. (2017). The *Neurozone Model of Brain Performance.* Retrieved from: https://neurozone.com/#our-products.
39 Kahneman, D. (2011). *Thinking, fast and slow.* New York: Macmillan, p35.
40 Adler, M., & Ziglio, E. (1996). *Gazing into the oracle: The Delphi method and its application to social policy and public health.* London: Jessica Kingsley Publishers.
41 Damasio, A. R. (1999). *The feeling of what happens: Body and emotion in the making of consciousness.* New York: Houghton Mifflin Harcourt.
42 Epstein, S. (1994). Integration of the cognitive and the psychodynamic unconscious. *American Psychologist, 49*(8), 709.
43 Grawe, K. (2007). *Neuropsychotherapy: How the neurosciences inform effective psychotherapy.* Retrieved from: https://psycnet.apa.org/doiLanding?doi=10.1037%2F0033-3204.44.1.118.
44 Kahneman, D. (2011). *Thinking, fast and slow.* New York: Macmillan.
45 Gordon, E. (2008). NeuroLeadership and Integrative Neuroscience: "it's about VALIDATION stupid!" *NeuroLeadership Journal, 1,* 71-80.

46   Lieberman, M.D. (2007). Social Cognitive Neuroscience: A Review of Core Processes. *Annual Review of Psychology, 58*(1), 259–289. https://doi.org/10.1146/annurev. psych.58.110405.085654

47   Ibid.

48   Ibid.

49   Damasio, A. R. (1999). *The feeling of what happens: Body and emotion in the making of consciousness*. New York: Houghton Mifflin Harcourt, p295.

50   Kandel, E. R. (2006). *In search of memory*. New York: W. W. Norton.

51   Ibid, p3.

52   Hanson, R. (2015). *Just One Thing: Pay Attention!* Retrieved from: https://greatergood. berkeley.edu/article/item/just_one_thing_pay_attention

53   Fehmi, L., & Robbins, J. (2008). *The open-focus brain: Harnessing the power of attention to heal mind and body*. Boston, MA: Shambhala Publications.

54   Starr, D. (2018). *Two psychologists followed 1000 New Zealanders for decades. Here's what they found about how childhood shapes later life*. Retrieved from https://www. sciencemag.org/news/2018/02/two-psychologists-followed-1000-new-zealanders- decades-here-s-what-they-found-about-how.

55   Peckham, A. (1999). Urban Dictionary. Retrieved August 25, 2019, from https://www. urbandictionary.com/define.php?term=The Urban Dictionary.

56   Fehmi, L., & Robbins, J. (2008). *The open-focus brain: Harnessing the power of attention to heal mind and body*. Boston, MA: Shambhala Publications.

57   Ibid.

58   BrainyQuote. (n.d.). *Albert Einstein quotes*. Retrieved from: https://www.brainyquote. com/quotes/albert_einstein_104873

59   Worringer, B., Langner, R., Koch, I., Eickhoff, S. B., Eickhoff, C. R., & Binkofski, F. C. (2019). Common and distinct neural correlates of dual-tasking and task-switching: a meta-analytic review and a neuro-cognitive processing model of human multitasking. *Brain Structure and Function, 224*(5), 1845–1869.

60   BrainyQuotes. (n.d.). *Herbert A. Simon quotes*. Retrieved from: https://www. brainyquote.com/quotes/herbert_a_simon_181919

61   Quote Investigator. (2014). *Charles Darwin Quotes*. Retrieved from: https:// quoteinvestigator.com/2014/05/04/adapt/

62   Hebb D. O. (1949). *The Organisation of Behaviour*. New York, NY: John Wiley & Sons.

63   Kandel, E. R., Schwartz, J. M., & Jessell, T. M. (2013). *Principles of Neural Science. McGraw-Hill editon* (5th ed.). New York: McGraw-Hill. https://doi. org/10.1036/0838577016.

64   Kriegler, S. (2008). Getting Out of Here – Now! (course notes). Pretoria: University of Pretoria.

65   Grawe, K. (2007). *Neuropsychotherapy: How the neurosciences inform effective psychotherapy*. London: Routledge.

66   Kleim, J. A., & Jones, T. A. (2008). Principles of experience-dependent neural plasticity: implications for rehabilitation after brain damage. *Journal of Speech, Language, and Hearing Research, 51*, 225–239.

67   Ibid.

68   Doidge, N. (2007). *The Brain that Changes Itself: Stories of Personal Triumph from the Frontiers of Brain Science*. London: Penguin.

69   Bressler, S. L., & Menon, V. (2010). Large-scale brain networks in cognition: emerging methods and principles. *Trends in Cognitive Sciences, 14*(6), 277–290. https://doi.org/10.1016/j.tics.2010.04.004

70   Ibid.

71   Ibid, p285.

72   Andrews-Hanna, J. R. (2012). The brain's default network and its adaptive role in internal mentation. *The Neuroscientist, 18*(3), 251–270.

73   Lieberman, M. D. (2013). *Social: Why our brains are wired to connect.* Oxford: Oxford University Press.

74   Ibid.

75   Boyatzis, Richard E. (2014). Possible Contributions to Leadership and Management Development From Neuroscience. *Academy of Management Learning & Education,* 300–303.

76   Boyatzis, R. E. (2014). Possible Contributions to Leadership and Management Development From Neuroscience. *Academy of Management Learning & Education,* 300–303.

77   Beaty, R. E., Kenett, Y. N., Christensen, A. P., Rosenberg, M. D., Benedek, M., Chen, Q., ... Kane, M. J. (2018). Robust prediction of individual creative ability from brain functional connectivity. *Proceedings of the National Academy of Sciences, 115*(5), 1087–1092.

78   Selye, H. (2013). *Stress in health and disease.* New York, NY: Butterworth-Heinemann.

79   Sapolsky, R. M. (2017). *Behave: The biology of humans at our best and worst.* New York, NY: Penguin.

80   Ibid.

81   Garland, E. L., Fredrickson, B., Kring, A. M., Johnson, D. P., Meyer, P. S., & Penn, D. L. (2010). Upward spirals of positive emotions counter downward spirals of negativity: Insights from the broaden-and-build theory and affective neuroscience on the treatment of emotion dysfunctions and deficits in psychopathology. *Clinical Psychology Review, 30*(7), 849–864.

82   Yerkes R. M. & Dodson J. D. (1908). The relation of strength of stimulus to rapidity of habit-formation. *Journal of Comparative Neurology and Psychology.* 18: 459–482. doi:10.1002/cne.920180503.

83   Arnsten, A. F. T. (2009). Stress signalling pathways that impair prefrontal cortex structure and function. *Nature Reviews Neuroscience, 10*(6), 410–422. https://doi.org/10.1038/nrn2648.Stress

84   GoodReads. (n.d.) *Matthew Walker Quotes.* Retrieved from: https://www.goodreads.com/quotes/9844227-the-best-bridge-between-despair-and-hope-is-a

85   Huffington, A. (2016). *The sleep revolution: transforming your life, one night at a time.* London: Penguin Random House UK.

86   Payne, J. D. (2011). Learning, memory, and sleep in humans. *Sleep Medicine Clinics, 6*(1), 15–30.

87   American Sleep Association. (n.d.). *About Sleep: Tips, Quotes and More.* Retrieved from: https://www.sleepassociation.org/about-sleep/

88   Hobson, J. A., & Pace-Schott, E. F. (2002). The cognitive neuroscience of sleep: neuronal systems, consciousness and learning. *Nature Reviews Neuroscience, 3*(9), 679.

89   Payne, J. D. (2011). Learning, memory, and sleep in humans. *Sleep Medicine Clinics*, *6*(1), 15–30.

90   Stickgold, R. (2005). Sleep-dependent memory consolidation. *Nature*, *437*(7063), 1272.

91   National Institute of Neurological Disorders and Stroke (NIH). (2019). *Brain Basics: Understanding Sleep*. Retrieved from: https://www.ninds.nih.gov/Disorders/Patient-Caregiver-Education/Understanding-Sleep

92   Payne, J. D., & Kensinger, E. A. (2010). Sleep's Role in the Consolidation of Emotional Episodic Memories. *Current Directions in Psychological Science*, *19*(5), 290–295.

93   Ibid.

94   Payne, J. D. & Nadel, L. (2004). *Sleep, dreams, and memory consolidation: the role of the stress hormone cortisol*. Retrieved from: https://www.ncbi.nlm.nih.gov/pmc/articles/PMC534695/

95   Tabibnia, G., & Radecki, D. (2018). Resilience training that can change the brain. *Consulting Psychology Journal: Practice and Research*, *70*(1), 59.

96   Johns, M. (n.d.). About the ESS. Retrieved from: https://epworthsleepinessscale.com/about-the-ess/

97   GoodReads. (n.d.). *Willam James quotes*. Retrieved from: https://www.goodreads.com/quotes/8935162-the-faculty-of-voluntarily-bringing-back-a-wandering-attention-over

98   Boyatzis, R E, & McKee, A. (2005). *Resonant Leadership: Renewing Yourself and Connecting with Others Through Mindfulness, Hope, and Compassion*. Boston: Harvard Business School Press, p9.

99   Good, D. J., Lyddy, C. J., Glomb, T. M., Bono, J. E., Brown, K. W., Duffy, M. K., Baer, R.A., Brewer, J.A. & Lazar, S. W. (2016). Contemplating mindfulness at work: An integrative review. *Journal of Management*, *42*(1), 114–142.

100   Farb, N. A. S., Segal, Z. V, Mayberg, H., Bean, J., Mckeon, D., Fatima, Z., & Anderson, A. K. (2007). Attending to the present: Mindfulness meditation reveals distinct neural modes of self-reference. *Social Cognitive and Affective Neuroscience*, *2*(4), 313–322. https://doi.org/10.1093/scan/nsm030

101   De Couck, M., Caers, R., Musch, L., Fliegauf, J., Giangreco, A., & Gidron, Y. (2019). How breathing can help you make better decisions: Two studies on the effects of breathing patterns on heart rate variability and decision-making in business cases. *International Journal of Psychophysiology, 139*:1-9.

102   Quotes. (n.d.). *Ralph Waldo Emerson quotes*. Retrieved from: https://www.quotes.net/quote/634

103   Jha, A. P., Morrison, A. B., Dainer-Best, J., Parker, S., Rostrup, N., & Stanley, E. A. (2015). Minds "at attention": Mindfulness training curbs attentional lapses in military cohorts. *PloS One*, *10*(2), e0116889.

104   BrainyQuote. (n.d.). *Leonard Cohen quotes*. Retrieved from: https://www.brainyquote.com/quotes/leonard_cohen_139113

105   Garland, E. L., Fredrickson, B., Kring, A. M., Johnson, D. P., Meyer, P. S., & Penn, D. L. (2010). Upward spirals of positive emotions counter downward spirals of negativity: Insights from the broaden-and-build theory and affective neuroscience on the treatment of emotion dysfunctions and deficits in psychopathology. *Clinical Psychology Review*, *30*(7), 849–864.

106   Wegbreit, E., Franconeri, S., & Beeman, M. (2015). Anxious mood narrows attention in feature space. *Cognition and Emotion*, *29*(4), 668–677.

107 Rowe, G., Hirsh, J. B., & Anderson, A. K. (2007). Positive affect increases the breadth of attentional selection. *Proceedings of the National Academy of Sciences, 104*(1), 383–388.

108 Vanlessen, N., De Raedt, R., Koster, E. H. W., & Pourtois, G. (2016). Happy heart, smiling eyes: a systematic review of positive mood effects on broadening of visuospatial attention. *Neuroscience & Biobehavioral Reviews, 68*, 816–837.

109 Payne, J. D. (2011). Learning, memory, and sleep in humans. *Sleep Medicine Clinics, 6*(1), 15–30.

110 Boyatzis, R. E. (2014). Possible Contributions to Leadership and Management Development From Neuroscience. *Academy of Management Learning & Education,* 300–303.

111 Ibid, p.1978.

112 Boyatzis, R. E, Smith, M. L., & Beveridge, A. J. (2013). Coaching with compassion: Inspiring health, well-being, and development in organizations. *The Journal of Applied Behavioral Science, 49*(2), 153–178.

113 Fredrickson, B. (2009). *Positivity: Groundbreaking Research Reveals How to Embrace the Hidden Strength of Positive Emotions. Overcome Negativity, and Thrive.* New York: Crown.

114 Barsade, S. G. (2002). The ripple effect: Emotional contagion and its influence on group behavior. *Administrative Science Quarterly, 47*(4), 644–675.

115 Demartini, J. F. (2002). *The breakthrough experience.* Carlsbad, California: Hay House, Inc.

116 Piaf, E. (1956). *Non, je ne regrette rien, French song composed by Charles Dumont, with lyrics by Michel Vaucaire.* Retrieved from: https://en.wikipedia.org/wiki/Non,_je_ne_regrette_rien

117 Goodreads. (n.d.). *Warren Buffett quotes.* Retrieved from: https://www.goodreads.com/quotes/9803116-you-will-continue-to-suffer-if-you-have-an-emotional

118 BrainyQuote. (n.d.). *William Arthur Ward quotes.* Retrieved from: https://www.brainyquote.com/quotes/william_arthur_ward_676240

119 Kini, P., Wong, J., McInnis, S., Gabana, N., & Brown, J. W. (2016). The effects of gratitude expression on neural activity. *NeuroImage, 128*, 1–10.

120 The Healed Tribe. (n.d.). *Self regulation through heart coherence.* Retrieved from: https://www.thehealedtribe.com/heart-coherence-and-resilience

121 BrainyQuote. (n.d.). *Anne Frank quotes.* Retrieved from: https://www.brainyquote.com/quotes/anne_frank_104183

122 Gu, S. S., Lillicrap, T., Turner, R. E., Ghahramani, Z., Schölkopf, B., & Levine, S. (2017). Interpolated policy gradient: Merging on-policy and off-policy gradient estimation for deep reinforcement learning. In *Advances in Neural Information Processing Systems* (pp. 3846–3855).

123 LeDoux, J. E. (2012). Evolution of Human Emotion. *Progress in Brain Research, 195*, 431–442. https://doi.org/10.1016/B978-0-444-53860-4.00021-0.EVOLUTION

124 Carter, R. (2014). *The human brain book: An illustrated guide to its structure, function, and disorders.* New York, NY: Penguin, p. 124.

125 Goldberg, R. (2001). *Performance art: From futurism to the present.* New York: Thames & Hudson.

126 Todorov, A. (2017). *Face value: The irresistible influence of first impressions.* Princeton, New Jersey: Princeton University Press.

127 LeDoux, J. E. (2012). Evolution of Human Emotion. *Progress in Brain Research, 195*, 431–442. https://doi.org/10.1016/B978-0-444-53860-4.00021-0.EVOLUTION

128 Ibid.

129 Ekman P. (1977). Biological and Cultural Contributions to Body and Facial Movement. In Blacking, J. (Ed.), *The Anthropology of the Body,* (pp. 34-84). London: Academic Press.

130 Ekman, P. (2019). *Atlas of Emotions.* Retrieved from http://atlasofemotions.org/

131 Ward, J. (2016). *The student's guide to social neuroscience.* New York: Psychology Press.

132 Lindquist, K. A., & Barrett, L. F. (2008). Constructing emotion: The experience of fear as a conceptual act. *Psychological Science, 19*(9), 898–903.

133 Russell, J. A., & Barrett, L. F. (1999). Core affect, prototypical emotional episodes, and other things called emotion: dissecting the elephant. *Journal of Personality and Social Psychology, 76*(5), 805.

134 Ibid.

135 Barrett, L. F., Adolphs, R., Marsella, S., Martinez, A. M., & Pollak, S. D. (2019). Emotional expressions reconsidered: challenges to inferring emotion from human facial movements. *Psychological Science in the Public Interest, 20*(1), 1–68.

136 Siegel, E. H., Sands, M. K., Van den Noortgate, W., Condon, P., Chang, Y., Dy, J., Quigley, K. S. & Barrett, L. F. (2018). Emotion fingerprints or emotion populations? A meta-analytic investigation of autonomic features of emotion categories. *Psychological Bulletin, 144*(4), 343.

137 Barrett, L. F., Adolphs, R., Marsella, S., Martinez, A. M., & Pollak, S. D. (2019). Emotional expressions reconsidered: challenges to inferring emotion from human facial movements. *Psychological Science in the Public Interest, 20*(1), 1–68.

138 Aldao, A., & Nolen-Hoeksema, S. (2010). Specificity of cognitive emotion regulation strategies: A transdiagnostic examination. *Behaviour Research and Therapy, 48*(10), 974–983.

139 Gross, J. J., & John, O. P. (2003). Individual differences in two emotion regulation processes: implications for affect, relationships, and well-being. *Journal of Personality and Social Psychology, 85*(2), 348.

140 Ibid.

141 Gross, J. J. (2007). *Handbook of emotion regulation.* New York: Guilford Publications.

142 Ibid.

143 Lindquist, K. A., & Barrett, L. F. (2008). Constructing emotion: The experience of fear as a conceptual act. *Psychological Science, 19*(9), 898–903.

144 Barrett, L. F. (2017). *How emotions are made: The secret life of the brain.* Boston, MA: Houghton Mifflin Harcourt.

145 Fosslien, L., & Duffy, M. W. (2019). *No Hard Feelings: The Secret Power of Embracing Emotions at Work.* New York: Penguin Random House.

146 Kashdan, T. B., Barrett, L. F., & McKnight, P. E. (2015). Unpacking emotion differentiation: Transforming unpleasant experience by perceiving distinctions in negativity. *Current Directions in Psychological Science, 24*(1), 10–16.

147 Goldin, P. R., McRae, K., Ramel, W., & Gross, J. J. (2008). The neural bases of emotion regulation: reappraisal and suppression of negative emotion. *Biological Psychiatry, 63*(6), 577–586.

148 Gross, J. J. (2008). Emotion regulation. *Handbook of Emotions, 3*(3), 497–513.

149 Killingsworth, M. A., & Gilbert, D. T. (2010). A wandering mind is an unhappy mind. *Science, 330*(6006), 932, p932.

150 Gross, J. J. (2007). *Handbook of emotion regulation*. New York: Guilford Publications.

151 Good, D. J., Lyddy, C. J., Glomb, T. M., Bono, J. E., Brown, K. W., Duffy, M. K., … Lazar, S. W. (2016). Contemplating Mindfulness at Work: An Integrative Review. *Journal of Management, 42*(1), 114–142. https://doi.org/10.1177/0149206315617003

152 Garland, E. L., Fredrickson, B., Kring, A. M., Johnson, D. P., Meyer, P. S., & Penn, D. L. (2010). Upward spirals of positive emotions counter downward spirals of negativity: Insights from the broaden-and-build theory and affective neuroscience on the treatment of emotion dysfunctions and deficits in psychopathology. *Clinical Psychology Review, 30*(7), 849–864.

153 International Network on Personal Meaning. (n.d.). *Paul T. P. Wong quote*. Retrieved from: https://www.latest.facebook.com/INPMeaning/posts/2716576068402850?__xts__[0]=68.ARC8tfmJMLTD5hUl7GQkFDO7ft38FLFqJKxP0CyL9S4BMpb5k6moA1c1HkJTvitnR-y-JzJFDrGKlEdTJWRLaizG_92Sk1F7Anq69ZbopGFjpHx5jiXT_G47__iWzTQeXX_ZeX_-wfUK7N448lMKyTof4UQn-b1i2eVjgm__i7QKYC4Vhno_focdsum7jztCXszbTNar7MnZbhMQSXUjN0YbHkP-y7zRerQf4cjKETgHLyu12JJr2FhbrojWVep5KoR4lVppvYbhc4oLRDXAmSOrrcejpN9prXGBS082jS4SgGqtQWKQtPaXBEkAxDje0316I7ulEN1hwApAA7kjADVNEw&__tn__=-R

154 Goodreads. (n.d.). *Jerry Seinfeld quotes*. Retrieved from: https://www.goodreads.com/quotes/162599-according-to-most-studies-people-s-number-one-fear-is-public

155 Adolphs, R. (2009). The Social Brain: Neural Basis of Social Knowledge. *Annual Review of Psychology, 60*, 693–716. https://doi.org/10.1146/annurev.psych.60.110707.163514.

156 Ochsner, K. N., & Lieberman, M. D. (2001). The emergence of social cognitive neuroscience. *American Psychologist, 56*(9), 717–734.

157 Dunbar, R. I. M. (1992). Neocortex Size as a Constraint on Group-Size in Primates. *Journal of Human Evolution, 22*(6), 469–493. https://doi.org/10.1016/0047-2484(92)90081-J

158 Tabibnia, G., & Radecki, D. (2018). Resilience training that can change the brain. *Consulting Psychology Journal: Practice and Research, 70*(1), 59, p67.

159 Bressler, S. L., & Menon, V. (2010). Large-scale brain networks in cognition : emerging methods and principles. *Trends in Cognitive Sciences, 14*(6), 277–290. https://doi.org/10.1016/j.tics.2010.04.004

160 Sporns, O. (2013). Structure and function of complex brain networks. *Dialogues in Clinical Neuroscience, 15*(3), 247–262. https://doi.org/10.1007/978-3-642-40308-8_2

161 Spunt, R. P., Meyer, M. L., & Lieberman, M. D. (2015). The default mode of human brain function primes the intentional stance. *Journal of Cognitive Neuroscience, 27*(6), 1116–1124.

162 Lieberman, M. D. (2013). *Social: Why our brains are wired to connect*. Oxford: Oxford University Press, p11.

163 Ibid.

164 Ibid.

165 Ibid.

166 Kandel, E. R., Schwartz, J. M., & Jessell, T. M., Siegelbaum S. A., Hudspeth A. J. & Kandel, M. S. (2013). *Principles of Neural Science*, (5th ed.). New York: McGraw-Hill. https://doi.org/10.1036/0838577016

167 Rossouw, J. G., & Rossouw, P. J. (2017). *Executive Resilience: Neuroscience for the Business of Disruption* (1st ed.). Sydney, Australia: Driven Books, p14.

168 Millner, D. (2013). *Jealousy (Part 2) – Six steps to overcoming jealousy*. Retrieved June 15, 2015, from https://www.eve.com.mt/2013/02/21/jealousy/

169 Lieberman, M. D. (2013). *Social: Why our brains are wired to connect*. Oxford: Oxford University Press, p60.

170 Ibid, p39.

171 Kross, E., Egner, T., Ochsner, K., Hirsch, J., & Downey, G. (2007). Neural dynamics of rejection sensitivity. *Journal of Cognitive Neuroscience, 19*(6), 945–956.

172 Kets de Vries, M. F. R. (1991). *Organizations on the couch: Clinical perspectives on organizational behavior and change*. San Francisco: Jossey-Bass.

173 Adolphs, R. (2009). The Social Brain: Neural Basis of Social Knowledge. *Annual Review of Psychology, 60*, 693–716. https://doi.org/10.1146/annurev.psych.60.110707.163514.

174 Cozolino, L. (2006). The social brain. *Psychotherapy in Australia, 12*(2), 12.

175 Bion, W. R. (1961). *Experiences in Groups*. New York: Basic Books.

176 Kets de Vries, M. F. R. (1991). *Organizations on the couch: Clinical perspectives on organizational behavior and change*. San Francisco: Jossey-Bass.

177 Cilliers, F., & Koortzen, P. (2000). The psychodynamic view on organisational behaviour. *The Industrial-Organizational Psychologist, 38*(2), 59–67.

178 Kets de Vries, M. F. R. (1991). *Organizations on the couch: Clinical perspectives on organizational behavior and change*. San Francisco: Jossey-Bass.

179 Cozolino, L. (2006). The social brain. *Psychotherapy in Australia, 12*(2), 12.

180 Ibid.

181 Kets de Vries, M. F. R. (1991). *Organizations on the couch: Clinical perspectives on organizational behavior and change*. San Francisco: Jossey-Bass.

182 Lieberman, M. D., & Eisenberger, N. I. (2008). The pains and pleasures of social life: A social cognitive neuroscience approach. *IN PRESS, Neuroleadership*, 1–38.

183 Ibid, p891.

184 Lieberman, M. D. (2013). *Social: Why our brains are wired to connect*. Oxford: Oxford University Press.

185 DeLong, T. J. (2011). *The Comparing Trap*. Retrieved from: https://hbr.org/2011/06/the-comparing-trap

186 Takahashi, H., Kato, M., Matsuura, M., Mobbs, D., Suhara, T., & Okubo, Y. (2009). When your gain is my pain and your pain is my gain: neural correlates of envy and schadenfreude. *Science, 323*(5916), 937–939.

187 Hart, S. L., & Legerstee, M. (2010). *Handbook of Jealousy: Theory, Research, and Multidisciplinary Approaches*. Malden: Wiley-Blackwell.

188 Holt-Lunstad, J., Smith, T. B., & Layton, J. B. (2010). Social relationships and mortality risk: a meta-analytic review. *PLoS Medicine, 7*(7), e1000316. https://doi.org/10.1371/journal.pmed.1000316

189 Zak, P. J. (2018). The Neuroscience of High-Trust Organizations. *Consulting Psychology Journal: Practice and Research, 70*(1), 45–58.

190 Demartini, J. F. (2013). *The Values Factor: The Secret to Creating an Inspired and Fulfilling Life*. New York: Berkley.

191 Glaser, J. E. (2014). *Conversational intelligence: How great leaders build trust and get extraordinary results* (1st ed.). UK: Routledge.

192 Zak, P. J. (2018). The Neuroscience of High-Trust Organizations. *Consulting Psychology Journal: Practice and Research, 70*(1), 45–58.

193 Rizzolatti, G. & Sinigaglia, C. (2008). *Mirrors in the Brain: How Our Minds Share Actions and Emotions.* Oxford, UK: Oxford University Press.

194 Cozolino, L. (2006). The social brain. *Psychotherapy in Australia, 12*(2), 12.

195 Rizzolatti, G., & Craighero, L. (2004). The mirror-neuron system. *Annual Review of Neuroscience, 27*, 169–192.

196 Lieberman, M. D., & Pfeifer, J. H. (2005). The self and social perception: Three kinds of questions in social cognitive neuroscience. In A Easton & N Emery. (2005). *The Cognitive Neuroscience of Social Behaviour.* London: Psychology Press, p. 195–235.

197 Emerson, R. W. (2004). *The Spiritual Emerson: Essential Writings.* Boston: Beacon Press.

198 Goleman, D., & Boyatzis, R. (2008). Social intelligence and the biology of leadership. *Harvard Business Review, 86*(9), 74–81.

199 Hughes, L. W. (2005). Developing transparent relationships through humor in the authentic leader-follower relationship. *Authentic Leadership Theory and Practice: Origins, Effects and Development, 3*, 83–106.

200 Bond, G. R. (2004). Supported employment: evidence for an evidence-based practice. *Psychiatric Rehabilitation Journal, 27*(4), 345.

201 Goldin, P. R., McRae, K., Ramel, W., & Gross, J. J. (2008). The neural bases of emotion regulation: reappraisal and suppression of negative emotion. *Biological Psychiatry, 63*(6), 577–586.

202 Sapolsky, R. M. (2017). *Behave: The biology of humans at our best and worst.* New York, NY: Penguin.

203 Frith, C. (2009). *Making up the mind: How the Brain Creates Our Mental World.* Hoboken, NJ: Wiley-Blackwell.

204 Dennett, D. C. (1988). Précis of the intentional stance. *Behavioral and Brain Sciences, 11*(3), 495–505.

205 Premack, D., & Woodruff, G. (1978). Does the chimpanzee have a theory of mind? *Behavioral and Brain Sciences, 1*(04), 515. https://doi.org/10.1017/S0140525X00076512

206 Jung, C. G. (2010). *Dreams: (From Volumes 4, 8, 12, and 16 of the Collected Works of CG Jung)(New in Paper)* (Vol. 1). Princeton, NJ: University Press.

207 Zaki, J. (2019). *The war for kindness: Building empathy in a fractured world.* New York: Crown Publishing Group.

208 Eisenberger, N. I. (2012). The pain of social disconnection: Examining the shared neural underpinnings of physical and social pain. *Nature Reviews Neuroscience, 13*(6), 421–434. https://doi.org/10.1038/nrn3231

209 Davis, M. H., Mitchell, K. V, Hall, J. A., Lothert, J., Snapp, T., & Meyer, M. (1999). Empathy, expectations, and situational preferences: Personality influences on the decision to participate in volunteer helping behaviors. *Journal of Personality, 67*(3), 469–503.

210 Eres, R., Decety, J., Louis, W. R., & Molenberghs, P. (2015). NeuroImage Individual differences in local gray matter density are associated with differences in affective and cognitive empathy. *NeuroImage, 117*, 305–310. https://doi.org/10.1016/j.neuroimage.2015.05.038

211 Sapolsky, R. M. (2017). *Behave: The biology of humans at our best and worst.* New York, NY: Penguin, p545.

212 Davis, M. H., Mitchell, K. V, Hall, J. A., Lothert, J., Snapp, T., & Meyer, M. (1999). Empathy, expectations, and situational preferences: Personality influences on the decision to participate in volunteer helping behaviors. *Journal of Personality, 67*(3), 469–503.

213 Bargh, J. A. (2006). *Social psychology and the unconscious: The automaticity of higher mental processes.* Philadelphia: Psychology Press.

214 Creative Mornings. (2019). *Weekly Highlights, quote by Yeshe Dawa.* Retrieved from: https://us2.campaign-archive.com/?u=b8d0c821758ac4979998d6301&id=d009ea4286 &e=%5BUNIQID%5D

215 Stout-Rostron, S. (2019). *Transformational Coaching to Lead Culturally Diverse Teams.* London: Routledge.

216 Cacioppo, J. T. (2002). Social neuroscience: Understanding the pieces fosters understanding the whole and vice7 versa. *American Psychologist, 57*(11), 819, p.820.

217 Tajfel, H. (1970). Experiments in intergroup discrimination. *Scientific American, 223*(5), 96–103.

218 Tajfel, H., Turner, J. C., Austin, W. G., & Worchel, S. (1979). An integrative theory of intergroup conflict. *Organizational Identity: A Reader*, 56–65.

219 Allport, G. W., Clark, K., & Pettigrew, T. (1954). The nature of prejudice. *International Society of Political Psychology, 12*(1), p 125-157.

220 Sapolsky, R. M. (2017). *Behave: The biology of humans at our best and worst.* New York, NY: Penguin.

221 Lieberman, M. D. (2013). *Social: Why our brains are wired to connect.* Oxford: OUP.

222 Singer, T., Seymour, B., O'doherty, J., Kaube, H., Dolan, R. J., & Frith, C. D. (2004). Empathy for pain involves the affective but not sensory components of pain. *Science, 303*(5661), 1157–1162.

223 Doyle, G., Srivastava, S. B., Goldberg, A., & Frank, M. C. (2017). Alignment at work: Using language to distinguish the internalization and self-regulation components of cultural fit in organizations. *ACL 2017 – 55th Annual Meeting of the Association for Computational Linguistics, Proceedings of the Conference (Long Papers), 1*, 603–612. https://doi.org/10.18653/v1/P17-1056

224 Rock, D., (2009). *Your Brain at Work.* New York: Harper Collins Publishers.

225 Sherman, S. M., Cheng, Y.-P., Fingerman, K. L., & Schnyer, D. M. (2016). Social support, stress and the aging brain. *Social Cognitive and Affective Neuroscience, 11*(7), 1050–1058.

226 Kahneman, D. (2011). *Thinking, fast and slow.* New York: Macmillan, p35.

227 Cunningham W.A., Zelazo, P. D., Packer, D. J. & van Bavel, J. J. (2007). The iterative reprocessing model: a multilevel framework for attitudes and evaluation. *Social Cognition, 25,* 736-60.

228 Haselton, M. G., Nettle, D., & Murray, D. R. (2015). The evolution of cognitive bias. *The Handbook of Evolutionary Psychology*, 1–20.

229 Kahneman, D. (2011). *Thinking, fast and slow.* New York: Macmillan, p35.

230 Greifeneder, R., Bless, H., & Fiedler, K. (2017). *Social cognition: How individuals construct social reality.* London: Psychology Press.

231 Gigerenzer, G. (1996). On Narrow Norms and Vague Heuristics : A Reply to Kahneman and Tversky (1996). *Psychological Review, 103*(3), 592–596.

232 Merriam-Webster Dictionary. (1928). Retrieved from https://www.merriam-webster.com/

233 VandenBos, G. R. (Ed.). (2007). *APA Dictionary of Psychology*. Washington, DC: American Psychological Association.

234 Telzer, E. H. (2016). Dopaminergic reward sensitivity can promote adolescent health: A new perspective on the mechanism of ventral striatum activation. *Developmental Cognitive Neuroscience, 17*, 57–67, p59. https://doi.org/10.1016/j.dcn.2015.10.010

235 Becker, W. J., Cropanzano, R., & Sanfey, A. G. (2011). Organizational Neuroscience: Taking organizational theory inside the neural black box. *Journal of Management, 37*(4), 933–961, p941. https://doi.org/10.1177/0149206311398955

236 Mitchell, J. P., Macrae, C. N., & Banaji, M. R. (2006). Dissociable medial prefrontal contributions to judgments of similar and dissimilar others. *Neuron, 50*(4), 655–663.

237 Molefi, N. (2017). *A Journey of Diversity and Inclusion in South Africa: Guidelines for Leading Inclusively*. Randburg, South Africa: Knowledge Resources Publishing.

238 Ibid.

239 The EW.Group. (2019). *What are micro-behaviours and how do they impact inclusive cultures?* Retrieved from: https://theewgroup.com/micro-behaviours-impact-inclusive-cultures/

240 Hobson, M. (2014). *Color blind or color brave?* Retrieved from: https://www.ted.com/talks/mellody_hobson_color_blind_or_color_brave?language=en

241 Spunt, R. P., Satpute, A. B., & Lieberman, M. D. (2011). Identifying the what, why, and how of an observed action: an fMRI study of mentalizing and mechanizing during action observation. *Journal of Cognitive Neuroscience, 23*(1), 63–74.

242 Ames, D. L., Jenkins, A. C., Banaji, M. R., & Mitchell, J. P. (2008). Taking Another Person's Perspective Increases Self-Referential Neural Processing. *Psychological Science, 19*(7), 642–644, p642. https://doi.org/10.1111/j.1467-9280.2008.02135.x

243 Goodreads. (n.d.). *Daniel Kahneman quotes*. Retrieved from: https://www.goodreads.com/quotes/885710-odd-as-it-may-seem-i-am-my-remembering-self

244 Rothmann, S. (2013). From happiness to flourishing at work: A Southern African perspective. In M. P. Wissing (Ed.), *Well-being research in South Africa* (pp. 123–151). Dordrecht: Springer.

245 Ibid

246 Kim, S., Reeve, J. & Bong, M. (2016a). *Recent Developments in Neuroscience Research on Human Motivation*. Bingley, UK: Emerald Publishing.

247 Goodreads. (n.d.). *Albert Einstein quotes*. Retrieved from: https://www.goodreads.com/quotes/423568-i-believe-in-intuition-and-inspiration-imagination-is-more-important

248 Alberts, I. (2018). *Passing the Torch: Preserving family wealth beyond the third generation*. (1st ed.). Hoboken: John Wiley & Sons.

249 Kounios, J., & Beeman, M. (2014). The cognitive neuroscience of insight. *Annual Review of Psychology, 65*(1), 71-93.

250 Fredrickson, B. (2009). *Positivity: Groundbreaking Research Reveals How to Embrace the Hidden Strength of Positive Emotions. Overcome Negativity, and Thrive*. New York: Crown, 2009.

251 Grawe, K. (2007). *Neuropsychotherapy: How the neurosciences inform effective psychotherapy*. London: Routledge.

252 Senge, M. O. (2006). Exercises in molecular gymnastics—bending, stretching and twisting porphyrins. *Chemical Communications*, (3), 243–256.

253 Bressler, S. L., & Menon, V. (2010). Large-scale brain networks in cognition: Emerging methods and principles. *Trends in Cognitive Sciences*, *14*(6), 277–290. https://doi.org/10.1016/j.tics.2010.04.004

254 Kandel, E. R., Schwartz, J. M., & Jessell, T. M., Siegelbaum S. A., Hudspeth A. J. & Kandel, M. S. (2013). *Principles of Neural Science*, (5th ed.). New York: McGraw-Hill, p1103 https://doi.org/10.1036/0838577016

255 Verschure, P. F. M. J., Pennartz, C. M. A., & Pezzulo, G. (2014). The why, what, where, when and how of goal-directed choice: neuronal and computational principles. *Philosophical Transactions of the Royal Society B: Biological Sciences*, *369*(1655), 20130483.

256 Berridge, K. C., Robinson, T. E., & Aldridge, J. W. (2009). Dissecting components of reward:'liking','wanting', and learning. *Current Opinion in Pharmacology*, *9*(1), 65–73.

257 Ibid.

258 Verschure, P. F. M. J., Pennartz, C. M. A., & Pezzulo, G. (2014). The why, what, where, when and how of goal-directed choice: neuronal and computational principles. *Philosophical Transactions of the Royal Society B: Biological Sciences*, *369*(1655), 20130483.

259 Kim, S., Reeve, J. & Bong, M. (2016a). Recent Developments in Neuroscience Research on Human Motivation. Bingley, UK: Emerald Publishing.

260 Frankl, V. (1959). *Man's search for meaning*. New York: Random House.

261 Ibid.

262 McKnight, P. E., & Kashdan, T. B. (2009). Purpose in life as a system that creates and sustains health and well-being: An integrative, testable theory. *Review of General Psychology*, *13*(3), 242–251, p. 242.

263 Rainey, L. (2014). *The search for purpose in life: An exploration of purpose, the search process, and purpose anxiety*. Pennsylvania: University of Pennsylvania, p22.

264 Baumeister, R. F., Vohs, K. D., Aaker, J. L., and Garbinsky, E. N. (2012). *Some Key Differences Between A Happy Life and a Meaningful Life*. Stanford, CA: Stanford University.

265 Elangovan, A. R., Pinder, C. C., & McLean, M. (2010). Callings and organizational behavior. *Journal of Vocational Behavior*, *76*(3), 428–440.

266 Seligman, M. E. P. (2012). *Flourish: A visionary new understanding of happiness and well-being*. New York: Simon and Schuster.

267 Duffy, R. D., & Sedlacek, W. E. (2007). The presence of and search for a calling: Connections to career development. *Journal of Vocational Behavior*, *70*(3), 590–601.

268 Dossey, L. (2018). The Helper's High. *EXPLORE*, *14*(6), 393–399.

269 Cutler, J., & Campbell-Meiklejohn, D. (2019). A comparative fMRI meta-analysis of altruistic and strategic decisions to give. *Neuroimage*, *184*, 227–241.

270 Frankl, V. E. (1985). *Man's search for meaning*. New York: Simon and Schuster.

271 Maslow, A. H. (1971). *The farther reaches of human nature*. New York: Viking Press, p269.

272 Steinmann, N. (2017). *Crucial Mentoring Conversations*, (1st ed.). Randburg, South Africa: KR Publishing.

273 Goodreads. (n.d.). *Kalu Ndukwe Kalu quites*. Retrieved from: https://www.goodreads.com/quotes/56602-the-things-you-do-for-yourself-are-gone-when-you

274 Uhl-Bien, M., & Arena, M. (2017). Complexity leadership: Enabling people and organizations for adaptability. *Organizational Dynamics*, *46*(1), 9–20.

275 Ryan, R. M., & Deci, E. L. (2000). Intrinsic and extrinsic motivations: Classic definitions and new directions. *Contemporary Educational Psychology, 25*(1), 54–67.

276 Boyatzis, Richard E, & Akrivou, K. (2006). The ideal self as the driver of intentional change. *Journal of Management Development, 25*(7), 624–642.

277 Ibid.

278 Boyatzis, R. E, Smith, M. L., & Van Oosten, E. (2009). Why coaching doesn't always work: coaching with compassion versus coaching for compliance. *Journal of Management Development, 25*(7), 624–602.

279 Uhl-Bien, M., & Arena, M. (2017). Complexity leadership: Enabling people and organizations for adaptability. *Organizational Dynamics.*

280 van Lill, X., Roodt, G. G., & de Bruin, G. P. (2019). Democratising Goal Setting: Possibilities and Pitfalls of Online Deliberation and Big Data Methods. In M. Coetzee, *Thriving in Digital Workspaces* (pp. 167–196). Springer Nature, Switzerland: Springer.

281 Fitzsimons, G. M., & Finkel, E. J. (2011). Outsourcing self-regulation. *Psychological Science, 22*(3), 369–375.

282 Gray, J. A. (1970). The psychophysiological basis of introversion-extraversion. *Behaviour Research and Therapy, 8*(3), 249–266.

283 Berkman, E. T. (2018). The neuroscience of goals and behavior change. *Consulting Psychology Journal: Practice and Research, 70*(1), 28.

284 Ibid.

285 Peters, J., & Büchel, C. (2010). Episodic future thinking reduces reward delay discounting through an enhancement of prefrontal-mediotemporal interactions. *Neuron, 66*(1), 138–148.

286 Ryan, R. M., & Deci, E. L. (2000). Intrinsic and extrinsic motivations: Classic definitions and new directions. *Contemporary Educational Psychology, 25*(1), 54–67.

287 Duckworth, A. L., & Seligman, M. E. P. (2006). Self-discipline gives girls the edge: Gender in self-discipline, grades, and achievement test scores. *Journal of Educational Psychology, 98*(1), 198–208. https://doi.org/10.1037/0022-0663.98.1.198

288 Rock, D. (2007). *Quiet Leadership: Six Steps to Transforming Performance at Work.* New York: Harper Collins Publisher.

289 van Lill, X., Roodt, G. G., & de Bruin, G. P. (2019). Democratising Goal Setting: Possibilities and Pitfalls of Online Deliberation and Big Data Methods. In M. Coetzee, *Thriving in Digital Workspaces* (pp. 167–196). Springer Nature, Switzerland: Springer.

290 MacLean, P. D. (1990). *The triune brain in evolution: Role in paleocerebral functions.* Berlin: Springer Science & Business Media.

291 Gordon, E., Barnett, K. J., Cooper, N. J., Tran, N., & Williams, L. M. (2008). An "Integrative Neuroscience" platform: application to profiles of negativity and positivity bias. *Journal of Integrative Neuroscience, 07*(03), 345–366. https://doi.org/10.1142/S0219635208001927

292 LeDoux, J. E. (2003). *Synaptic self: How our brains become who we are.* New York: Penguin.

293 LeDoux, J. E. (2012). Evolution of Human Emotion. *Progress in Brain Research., 195,* 431–442. https://doi.org/10.1016/B978-0-444-53860-4.00021-0.EVOLUTION

294 LeDoux, J. E. (2003). *Synaptic self: How our brains become who we are.* New York: Penguin.

295 Gordon, E. (2008). NeuroLeadership and Integrative Neuroscience: "it's about VALIDATION stupid!" *NeuroLeadership Journal, 1,* 71-80.

296 Kandel, E. R. (1998). A new intellectual framework for psychiatry. *American Journal of Psychiatry, 155*(4):457–469.

297 Kandel, E. R. (2005). Genes, brains and Self-understanding. Biology's aspirations for a new humanism. In: Kandel, E.R. (2005). *Psychiatry, Psychoanalysis and the new Biology of Mind*. Washington: American Psychiatric Publishing.

298 Rossouw, P. J. (2013). The neuroscience of talking therapies. Implications for therapeutic practice. *The Australian Journal of Counselling Psychology, 13*(1), 40-50.

299 Grawe, K. (2007). Neuropsychotherapy: How the neurosciences inform effective psychotherapy. *Psychotherapy: Theory, Research, Practice, Training, 44*(1), 118-120. https://doi.org/10.1037/0033-3204.44.1.118.

300 Ghadiri, A., Habermacher, A., & Peters, T. (2013). *Neuroleadership: A journey through the brain for business leaders*. Heidelberg, Germany: Springer Science & Business Media, p77.

301 Rossouw, J. G., & Rossouw, P. J. (2017). *Executive Resilience: Neuroscience for the Business of Disruption* (1st ed.). Sydney, Australia: Driven Books.

302 Weinberg, I. (2018). *Neurosurgeon Probes Wellness and Performance*. Retrieved from: http://ndabaonline.ukzn.ac.za/NewsletterPrinter.aspx?id=25

303 Frankl, V. E. (1985). *Man's search for meaning*. New York: Simon and Schuster.

304 Ader, R., & Cohen, N. (1975). Behaviorally conditioned immunosuppression. *Psychosomatic Medicine, 37*(4), 333–340.

305 Irwin, M. R. (2008). Human psychoneuroimmunology: 20 years of discovery. *Brain, Behavior, and Immunity, 22*(2), 129–139.

306 Ibid.

307 Weinberg, I. (2007). *The last frontier*. Johannesburg: Interpak Books.

308 Weinberg, I. (2013). *Engaging the Neural Matrix: Accessing the Multiple Dimensions of Consciousness*, p21. Retrieved from www.quantumpni.co.za

309 Van der Walt, E. (2017). The *Neurozone Model of Brain Performance*. Retrieved from: https://neurozone.com/#our-products

310 Ibid.

311 Ibid.

312 Ibid.

313 Ibid.

314 Ibid.

315 Rock, D. (2009). *Your Brain at Work*. New York: Harper Collins Publishers.

316 Arnsten, A. F. T. (2009). Stress signalling pathways that impair prefrontal cortex structure and function. *Nature Reviews Neuroscience, 10*(6), 410–422. https://doi.org/10.1038/nrn2648.Stress

317 Rock, D., & Tang, Y. (2009). Neuroscience of engagement. *Neuroleadership Journal, 2*, 1-8.

318 Radecki, D., Hull, L., McCusker, J., & Ancona, C. (2018). *Psychological Safety: The key to happy, high performing people and teams*. San Jose, CA: Red Hill Publishing.

319 Ibid.

Index